PROGRAMMED PRINCIPLES OF STATICS

PROGRAMMED PRINCIPLES OF STATICS

Gale E. Nevill, Jr.

Department of Engineering Science and Mechanics
University of Florida

JOHN WILEY & SONS, INC.,

NEW YORK · LONDON · SYDNEY · TORONTO

10 9 8 7 6 5 4 3 2 1
Library of Congress Catalog Card Number: 69-19927
SBN 471 63270 8
Printed in the United States of America

Preface

This book has many uses. It provides an effective introduction through independent study to the concepts and techniques of statics. For students enrolled in an introductory engineering mechanics course it will serve as a quick, efficient review or as supplementary help with the hard spots. It can also serve in physics courses as an introduction to the analysis of engineering problems. Instructors will find that its use for self-study brings their classes to a common level of understanding and relieves some of the pressure on classroom time. Also it helps the student to bridge the gap between classroom solution and independent solution of problems.

The coverage is similar to most one-semester engineering courses in statics and overlaps the widely adopted introductory texts covering mechanics. A knowledge of algebra is sufficient for understanding most of the material, although vectors and elementary calculus are used when appropriate.

The book is written in an informal, tutorial style that employs a programmed instruction format. It provides frequent opportunities to stop, think, and answer specific questions. Maximum learning results when the student writes his own answers in the spaces provided before he looks at the ones that are given. The answer column can be covered with the enclosed masking card and, since responses will be written in the book, the margins may be used for diagrams and calculations.

I hope that both the students and their instructors will find this approach of value in the understanding of statics.

Gale E. Nevill, Jr.

Contents

Chapter 1 Introduction 1

Definition of the field; principal physical quantities; principal idealizations and assumptions; Newton's Laws; and physical dimensions—unit changes

Chapter 2 Vectors as Tools 7

Basic nature of vector quantities; fundamental vector operations; resolution of vectors into components; use of rectangular components; unit vectors; alternative descriptions of vectors; scalar product of two vectors; and applications of scalar products

Chapter 3 Moments of Forces 21

General concept of the moment of a force; vector product of two vectors; use of cartesian components in vector product computation; moment of a force about a point—vector product definition; moment of a force about a point using force components; moment of a force about a line—vector and scalar approaches; and couples

Chapter 4 Force Systems 39

4.1 General Force Systems 39

Equivalence of two systems of forces; replacement of force by force and couple at different locations; and resultant of general force system

4.2 Some Special Force Systems 42

Concurrent; coplanar; and parallel

4.3 Distributed Force Systems 51

Chapter 5 First Moments—Centroids, Centers of Gravity, and Mass 57

Center of gravity of collection of particles; center of gravity of continuous bodies; center of mass/centroid of volume; centroid of areas; symmetry considerations; and composite area and volume approach

Chapter 6 Equilibrium 72

Basic concepts of equilibrium of particles and rigid bodies; equations of equilibrium: particles, rigid bodies in two dimensions, three-dimensional bodies; free body diagrams; general solution procedure for problems of equilibrium; special cases—two, three force bodies; equilibrium of composite systems; and equilibrium of sections of continuous bodies—internal forces

Chapter 7 Introduction to Structural Analysis 93

7.1 Trusses and Frames 93

Definition and characteristics of trusses; limitation to simple trusses; techniques of solution of simple plane trusses: equilibrium of joints—special loading conditions, equilibrium of sections of trusses; space trusses; frames as distinguished from trusses; and frames as problems of equilibrium of multicomponent systems

7.2 Beams 107

Shearing force and bending moment in beams; relations between moment, shear, and load; construction of shear and moment diagrams

7.3 Flexible Cables 116

Characteristics and uses of cables; cables with concentrated loads; general relations for plane cables with distributed loads; and parabolic cables

Chapter 8 Friction 129

Basic nature of friction; friction "laws"—static and dynamic—finite areas; classification of friction problems; general approach to the solution of problems with friction; wedges and square threaded screws; and belt friction

Chapter 9 Second Moments—Moments of Inertia 158

9.1 Second Moments of Areas 158

Definition and calculation by direct integration; polar second moments; parallel axis theorem; and composite area approach

9.2 Second Moments of Mass—Moments of Inertia 165

Definition and calculation by direct integration; parallel axis theorem; thin plates as a special case; and composite body approach

Chapter 10 Virtual Work 173

Some definitions; principle of virtual work for rigid bodies; principle of virtual work for connected system of rigid bodies; and use of virtual work in determining reactions

Index 183

PROGRAMMED PRINCIPLES OF STATICS

chapter 1

Introduction

Objectives

Our first task is to identify the types of engineering problems associated with statics and to introduce the principal physical quantities, idealizations, basic laws, and concepts of the field. Upon completion of this chapter, the student should be able to:

1. Examine simple engineering problems and correctly determine if they are problems of statics.
2. Name and define the three principal physical quantities of statics.
3. State the conditions under which the idealizations particle, point force, and rigid body are valid and give examples.
4. State Newton's laws of motion and of gravitation and explain why statics is not concerned with the second law of motion.
5. Determine if the units of a problem are consistent and correctly change from one set of units to another.

Before we can solve any significant problems in statics, we must lay a foundation of basic concepts and ideas. The physical world abounds with mechanical systems subjected to loads and experiencing deformations and motions. The general study of these systems is called mechanics. Statics is that division of mechanics dealing with the analysis of mechanical systems that are at rest or move with constant velocity. Thus, statics is the study of mechanical systems for which all parts have zero ________. acceleration

An important characteristic of statics is that it has as its foundation a few basic laws or concepts. These laws relate a small number of physical quantities. In statics we are primarily concerned with the determination of distributions of loads or forces in systems that are in equilibrium, that is, they have no acceleration. Thus, force is one of the principal physical quantities of interest in statics. The other physical quantities involved in mechanics are *mass*, the measure of the inertial property of matter; *weight*, the force gravity exerts on mass; *length*, a measure of size or distance; and *time*, which allows the ordering of events that occur. These quantities are defined in all standard texts. Since statics is the study of mechanical systems at rest or moving without acceleration, the more significant of these quantities here are

weight length

____ and ____. Ordinarily, mass and time are not significant in statics.

It is useful to talk of these physical quantities in statics in terms of idealizations, as we discuss the basic laws. These useful idealizations are particle, point force, and rigid body. We define a *particle* as an element of matter that has mass but whose dimensions are negligibly small compared to the other dimensions of the system. Thus we consider a pin

particle

larger

head or the sun as a ____ provided the other dimensions of the system are much ____ than the dimensions of these bodies.

Next, consider a *point force:* a force that is considered to act at precisely one point in space. In fact, all forces are distributed over some finite area or volume of matter. However, if the size of the area or volume involved is negligibly small compared with other system dimensions, then we consider the force

point

to act at a ____.

Another important idealization is the *rigid body.* By definition, the distances between all particles in a rigid body are constant. Actually, real bodies deform to some extent under load. Yet, if the deformation is negligible compared to other dimensions of interest,

rigid

we treat the body as ____.

It is important to remember that the validity of these idealizations depends on *relative* magnitudes. Thus, for a quantity to be neglected, its size must be small *compared to some other part of the system.*

We now examine some basic laws or rules of mechanics. Since these basic laws are few in number and are used to solve a wide variety of problems, it is worthwhile to learn them well. We shall need to know Newton's Laws of Motion. The first law states that *a particle will remain at rest or move on a straight line with constant velocity if there are no unbalanced forces acting on it.* The converse of this statement is also true. That is, a particle at rest has no ________ ______ acting on it. This first law states the inertial property of matter.

unbalanced forces

Newton's second law states that *the acceleration of a particle is proportional in magnitude to the sum of the unbalanced forces acting on it and is in the direction of this sum.* This law predicts how particles will behave when subjected to unbalanced forces. When the unbalanced forces are zero, the________ of the particle is zero, and the first law applies.

acceleration

Newton's third law states that *two bodies in contact exert forces on each other that are equal in magnitude and opposite in direction.* This law is often stated as the requirement that action and ______ be equal.

reaction

Which of the three laws will we *not* be concerned with in this study? __________. Why? ______________ ____.

The second law because statics involves systems at rest or moving with constant velocity.

Since Newton's laws are written for particles, how do we treat entire, continuous bodies? The answer is that we sum or add together the effects of the forces acting on the particles involved to reach conclusions about the behavior of the entire body. In this regard, we consider continuous bodies simply as collections of ______.

particles

Another part of the foundation of statics is Newton's law of gravitation. It states that *there is a force of attraction, F, between two particles directly proportional to the product of the masses, m_1, m_2, of the two particles and inversely proportional to the square of the distance, r, between them.* The formula may be written as $F = Gm_1m_2/r^2$ where G is called the gravitational constant. From this physical law we see that, as the masses of the particles increase, the gravitational force ________________ and, as
increases/decreases
masses move farther apart, the gravitational force
________________.
increases/decreases

increases

decreases

Next, we discuss units. A consistent set of units for physical quantities must be used throughout any problem. That is, distance should not be expressed in inches in one part of the problem and in feet in another. Since the standard units are defined in most texts, we will consider only how to change units from one system to another. The way to change units is to multiply by *unity* and, therefore, not change the value of the quantity involved. Let us demonstrate this with an example.

Suppose that you must change the units of a speed $V = 44$ ft/sec into an equivalent value in miles per hour. Here it is obvious that we must change feet into ____ and seconds into ____.

miles hours

To do this, use the following *unit value* fractions: $1 = 1$ mile/5280 ft and $1 = 3600$ sec/1 hr. Clearly we may multiply our speed by either of these fractions and not change its real value. Do this and obtain the following relation:

$$V = \frac{44 \text{ ft}}{\text{sec}} \times \frac{1 \text{ mile}}{5280 \text{ ft}} \times \frac{3600 \text{ sec}}{1 \text{ hr}}.$$

We have common factors of feet and seconds which may be cancelled from both numerator and denominator. Our result is (use slide rule) $V =$ ____ mph.

30

This simple example demonstrates that changing units correctly is a straightforward process as long as one remembers to multiply by fractions that equal ________.

one or unity

Finally, it is worth repeating that our view of statics if primarily concerned with forces that act on rigid bodies and particles. Furthermore, statics is concerned only with systems that are at rest or in motion with constant velocity.

Summary

Statics is the study of mechanical systems that have zero acceleration, that is, systems that are at rest or move with constant velocity. The principal physical quantities of statics are force, weight, and length. In order to analyze real systems, we often make idealizations. Principally we consider a *particle*, an element of mass with negligible size, a *point force* which is a force treated as acting at a point, and a *rigid body*, a body with negligible deformation or distortion. Newton's laws provide the principal tools of statics. In particular, the first law—that a particle with zero acceleration has zero force on it—and the third law—that two bodies in contact exert an equal and opposite force on each other—are important. In every problem, a consistent set of units must be used; to change units, multiply by fractions with value one.

Problems

(1) A man fishing with a 40-lb test line hauls a tire weighing 21 lb up from the water surface to a pier. If he reels at constant velocity (no acceleration), what force does the tire exert on the line? What force does the man exert on the line? Will the line break?

(2) A man wishes to lift a 200-lb engine block with a block and tackle. What force must he exert to lift the block with no acceleration when he uses each of the configurations shown in Fig. P1.1?

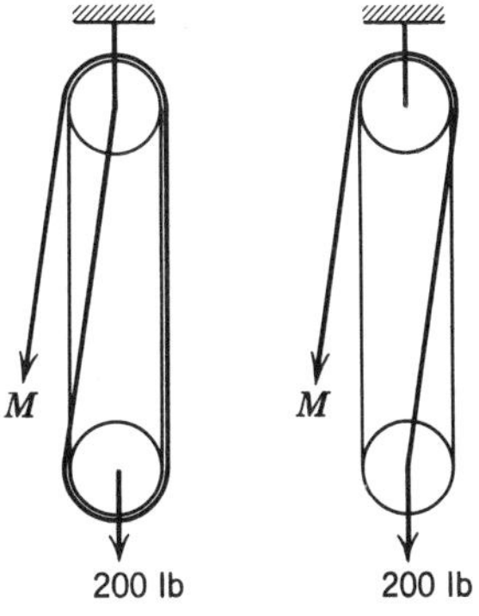

Figure P1.1

(3) In the arrangement shown in Fig. P1.2, does the spring scale read 0, 30 lb, or 60 lb?

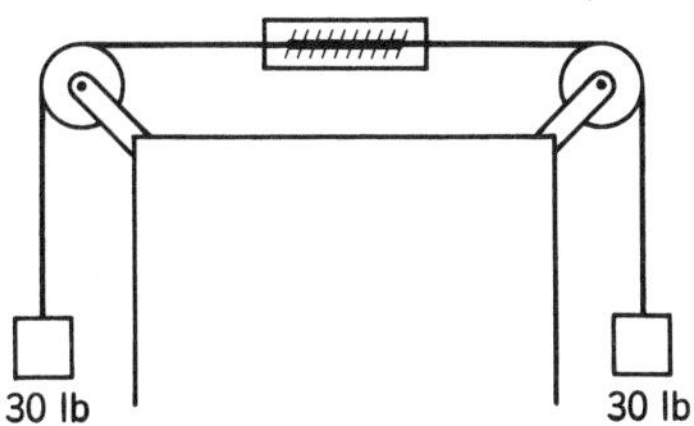

Figure P1.2

(4) If the speed of sound in air is 1,087 feet per second, what is its speed in miles per hour?

Answers to Problems

Chapter 1

(1) a. 21 lb
b. 21 lb
c. no

(2) a. 50 lb
b. 66.7 lb

(3) 30 lb

(4) 741 mph

chapter 2

Vectors as Tools

Objectives

In statics, and indeed throughout engineering and science, vector quantities are important. In this chapter we review some basic vector properties and operations, and examine various representations of vector quantities. After completing this chapter, the student should be able to:

1. State the characteristics of vectors, give examples of vector physical quantities, and define and illustrate the operations of addition and multiplication by a scalar.
2. Resolve vectors into orthogonal components and express vectors in terms of orthogonal unit vectors.
3. State three distinct vector representations and convert from one to another readily.
4. Define the vector scalar product, state its properties, and utilize it to solve for projections and angles.

Since the study of statics relies heavily upon the use of vectors, we begin with some mathematical preliminaries. It is convenient to contrast the behavior of vectors with that of *scalars*. We are familiar with the *scalar* quantities—mass, time, volume, and density: they are characterized by a single number that expresses their magnitude. They are added as ordinary algebraic numbers. *Vector* quantities, such as force, velocity, and displacement, are characterized by a magnitude, a characteristic direction, *and* a special rule for their addition.

We introduce a convenient symbol, the arrow, as a graphical representation of a vector. The arrow in Fig. 2.1 represents the vector $\bar{\mathbf{V}}$. The magnitude of

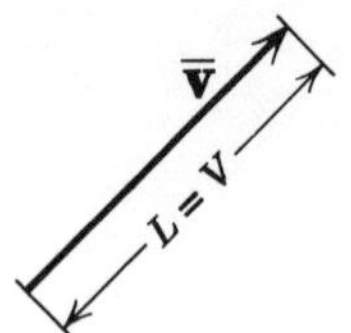

Figure 2.1

this vector is denoted in the figure by the symbol *L*.

Our notation will be to write the magnitude of a vector, $\bar{\mathbf{V}}$, as the letter *V*, without the bar. This vector magnitude, *V*, remains constant provided the arrow length is fixed, regardless of its orientation in space. The *magnitude* of a vector is not a vector quantity but a ____ quantity.

scalar

Thus the magnitude of a vector quantity is simply the size of that quantity. It is always positive in sign and is independent of the ____________ of the vector.

direction/orientation

There are a number of ways to describe the orientation or direction of a vector. The orientation of a vector may be described (see Fig. 2.2) by specifying

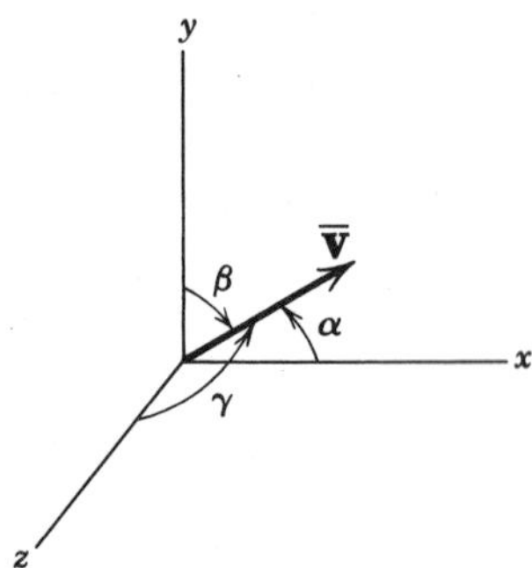

Figure 2.2

the angles between the vector and each of a chosen set of coordinate axes. In space, three angles are required to specify the direction uniquely. In two-dimensional or plane space, how many angles are required? ____.

One

Orientation can also be specified by two points

(Fig. 2.3). The vector $\overline{\mathbf{V}}$ is directed by stating that it is parallel to a line through points *A* and *B*. It is impor-

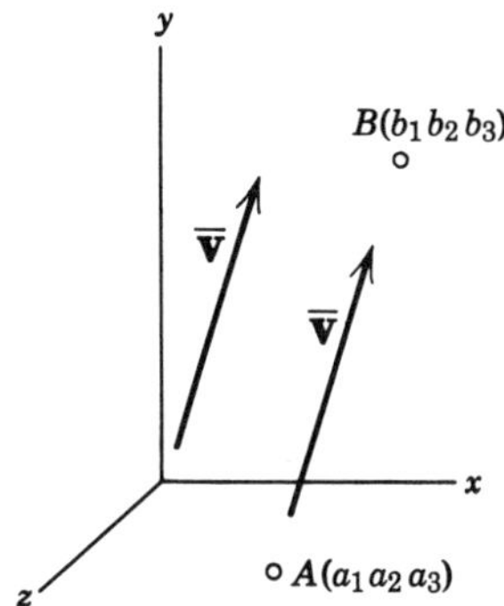

Figure 2.3

tant to note that the arrow representing the vector $\overline{\mathbf{V}}$ may be either longer or shorter than the distance between points *A* and *B*. Don't confuse the vector magnitude with the distance between points *A* and *B*. The positive direction (sense) of $\overline{\mathbf{V}}$ is prescribed by ordering the points. Thus, specification of an ordered pair of points gives the orientation and sense of a vector, but not its ________. magnitude

We next consider the multiplication of a vector by a scalar. By definition, the product $n\overline{\mathbf{V}}$, where n is a scalar and $\overline{\mathbf{V}}$ is a vector, is a vector with the same direction and sense as $\overline{\mathbf{V}}$ but with a magnitude n times the magnitude of $\overline{\mathbf{V}}$. Thus, if $\overline{\mathbf{V}}$ is a vector of length 2 directed from *A* to *B*, the vector $n\overline{\mathbf{V}}$, where $n = 5$, would be a vector of magnitude __ directed from __ to __. 10 *A* *B*

Now, what is the meaning of a negative sign associated with a vector? We define the vector $-\overline{\mathbf{V}}$ to be a vector with the same orientation and magnitude as $\overline{\mathbf{V}}$ but with opposite sense, i.e., in the opposite direction. Thus, if $\overline{\mathbf{V}}$ is directed from *A* to *B*, $-\overline{\mathbf{V}}$ (representing $\overline{\mathbf{V}}$ multiplied by -1) is directed from __ to __. *B* *A*

Multiplication of a vector by a negative scalar, therefore, changes both the magnitude and sense of the vector but not its ________. orientation

For example, if $\overline{\mathbf{V}}$ has magnitude 3 and is directed from A to B, the vector $n\bar{\mathbf{V}}$, when $n = -3$, has magnitude __ and is directed from __ to __.

9 B A

We next consider the meaning of vector equality. Two vectors are equal if they have the *same magnitude, orientation,* and *sense.* Thus, two vectors may be equal without having the same line of action.

Next consider the addition of vectors. This is defined by the parallelogram rule. In Fig. 2.4 we show

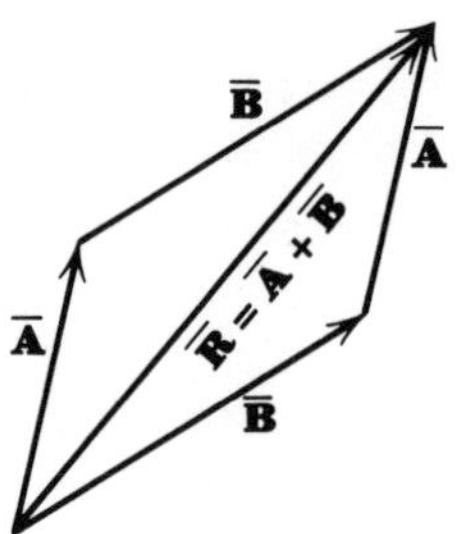

Figure 2.4

the sum of the two vectors $\bar{\mathbf{A}}$ and $\bar{\mathbf{B}}$. Since $\bar{\mathbf{A}}$ and $\bar{\mathbf{B}}$ define a plane in space, the resultant $\bar{\mathbf{R}} = \bar{\mathbf{A}} + \bar{\mathbf{B}}$ __________ lie in this plane.

will/will not

will

Will the magnitude of the resultant $\bar{\mathbf{R}} = \bar{\mathbf{A}} + \bar{\mathbf{B}}$ necessarily equal the sum of the magnitudes of $\bar{\mathbf{A}}$ and $\bar{\mathbf{B}}$? __.

No

Furthermore, vector addition *is commutative.* That is, $\bar{\mathbf{A}} + \bar{\mathbf{B}}$ ________ equal to $\bar{\mathbf{B}} + \bar{\mathbf{A}}$.

is/is not

is

When more than two vectors are added, it is also useful to use the *associative* property of vector addition. Thus, when vectors $\bar{\mathbf{A}}$, $\bar{\mathbf{B}}$, and $\bar{\mathbf{C}}$ are added, it does not matter whether $\bar{\mathbf{A}}$ and $\bar{\mathbf{B}}$ are added and their resultant is added to $\bar{\mathbf{C}}$ or whether some other order is chosen. We note also the *distributive* property of scalar multiplication. The sum of the vectors $n\bar{\mathbf{A}}$ and $n\bar{\mathbf{B}}$ is the same as n times the vector $(\bar{\mathbf{A}} + \bar{\mathbf{B}})$, i.e., $n\bar{\mathbf{A}} + n\bar{\mathbf{B}} = n(\bar{\mathbf{A}} + \bar{\mathbf{B}})$. From the above rules, could we conclude that $n\bar{\mathbf{A}} + n\bar{\mathbf{B}} + n\bar{\mathbf{C}}$ is equal to $n(\bar{\mathbf{C}}+\bar{\mathbf{B}}+\bar{\mathbf{A}})$? ____.

Yes

Now let's consider the resolution of vectors into component parts. We wish to determine a set of component vectors, with certain orientations, that add to give the original vector. We might resolve a vector into components in three *skew* directions as shown in Fig. 2.5. We seek components of the vector

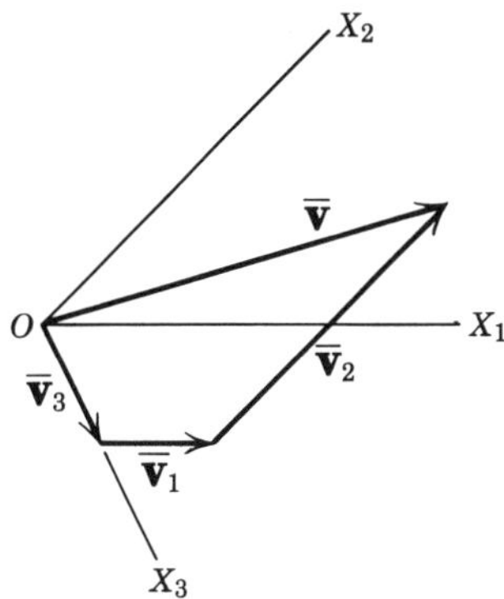

Figure 2.5

$\overline{\mathbf{V}}$ in the directions OX_1, OX_2, and OX_3, where these directions are neither parallel nor perpendicular to each other. From Fig. 2.5, by inspection, we can write $\overline{\mathbf{V}} = \_ + \_ + \_$ where these component vectors are each ____________ to the corresponding co- ordinate directions.
parallel/perpendicular

$\overline{\mathbf{V}}_1$ $\overline{\mathbf{V}}_2$ $\overline{\mathbf{V}}_3$
parallel

If the orientations of the coordinate axes and the vector $\overline{\mathbf{V}}$ are given, then it is straightforward, though time consuming, to find these components. Skew coordinates, however, are relatively uncommon because of the convenience and utility of rectangular (orthogonal) coordinates. By orthogonal we mean a set of *mutually perpendicular* coordinate directions (Fig. 2.6). Here we see that the orientation of $\overline{\mathbf{V}}$ relative to the set of orthogonal coordinate directions Ox, Oy, and Oz may be specified in terms of the angles, α, β, and γ. Also, the vector $\overline{\mathbf{V}}$ equals the sum of $\overline{\mathbf{V}}_x$, _, and _.

$\overline{\mathbf{V}}_y$ $\overline{\mathbf{V}}_z$

Note that the components $\overline{\mathbf{V}}_x$, $\overline{\mathbf{V}}_y$, and $\overline{\mathbf{V}}_z$ form the edges of a rectangular parallelepiped. Therefore, the square of the magnitude of $\overline{\mathbf{V}}$, (V^2), equals the sum of the squares of the magnitudes of $\overline{\mathbf{V}}_x$, _ and _.

$\overline{\mathbf{V}}_y$ $\overline{\mathbf{V}}_z$

We also see that $V_x = V \cos$ _, $V_y = V \cos$ _, and $V_z = V \cos$ _.

α β
γ

Finally, recall that $\cos^2\alpha + \cos^2\beta + \cos^2\gamma =$ ____.

one

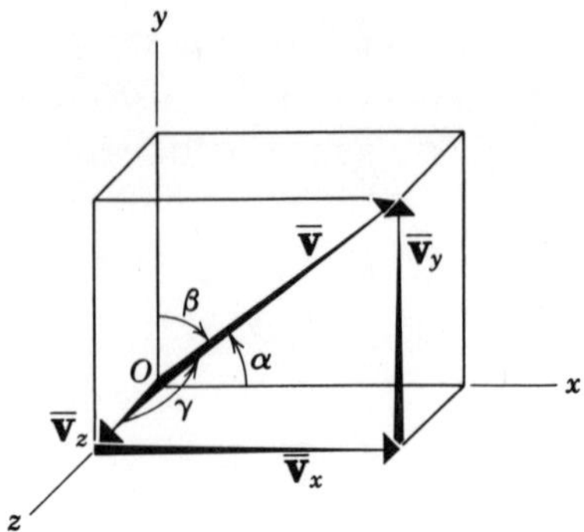

Figure 2.6

Using scratch paper, verify the consistency of these formulas by writing the components $V_x = V \cos \alpha$, $V_y = V \cos \beta$, and $V_z = V \cos \gamma$ and summing their squares.

For the two dimensional case (Fig. 2.7) where $\overline{\mathbf{V}}$

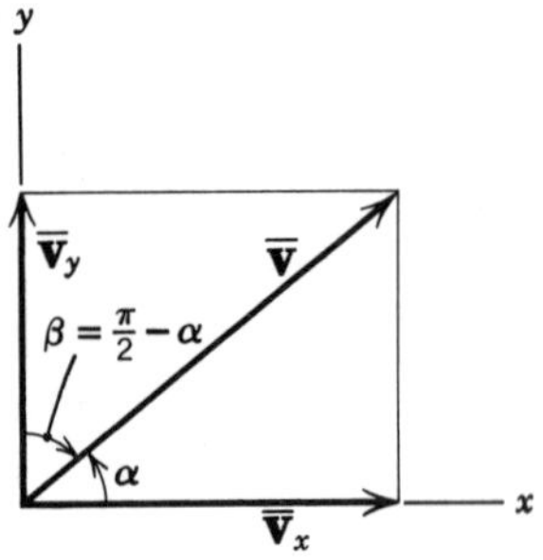

Figure 2.7

lies in one of the coordinate planes, only one angle
α (and its complement) is involved. Here $V_x = V \cos$ __
β α β and $V_y = V \cos$ __. Also $V_y = V \sin$ __ and $V_x = V \sin$ __.

To use the concept of components, we define the unit (length) vectors $\bar{\mathbf{i}}$, $\bar{\mathbf{j}}$, and $\bar{\mathbf{k}}$ in the *x*, *y*, and *z* directions, respectively, (Fig. 2.8). Let us express the vectors $\overline{\mathbf{V}}_x$, $\overline{\mathbf{V}}_y$, and $\overline{\mathbf{V}}_z$ in terms of $\bar{\mathbf{i}}$, $\bar{\mathbf{j}}$, and $\bar{\mathbf{k}}$. Consider $\overline{\mathbf{V}}_x$ first: since $\overline{\mathbf{V}}_x$ and $\bar{\mathbf{i}}$ are parallel and the magnitude of $\bar{\mathbf{i}}$ is 1, write $\overline{\mathbf{V}}_x$ (vector) $= V_x$ (magnitude) times $\bar{\mathbf{i}}$.
V_y $V_z\bar{\mathbf{k}}$ In a similar fashion, we write $\overline{\mathbf{V}}_y =$ __$\bar{\mathbf{j}}$ and $\overline{\mathbf{V}}_z =$ __.

The convenience of a unit vector description is

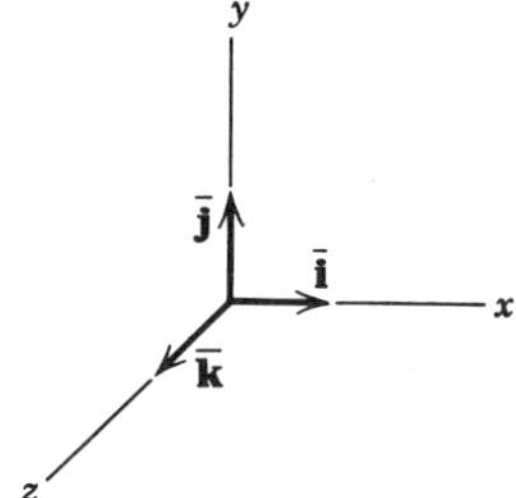

Figure 2.8

seen in the addition of several vectors. Consider adding $\bar{\mathbf{U}}$, $\bar{\mathbf{V}}$, and $\bar{\mathbf{W}}$, general three dimensional vectors. Instead of using the trigonometry of the parallelogram rule directly, let us express each of these vectors in component form as $\bar{\mathbf{V}} = V_x\bar{\mathbf{i}} + V_y\bar{\mathbf{j}} + V_z\bar{\mathbf{k}}$, $\bar{\mathbf{U}} = U_x\bar{\mathbf{i}} +$ __$\bar{\mathbf{j}} +$ ___, and $\bar{\mathbf{W}} =$ ______________.

U_y $U_z\bar{\mathbf{k}}$
$W_x\bar{\mathbf{i}} + W_y\bar{\mathbf{j}} + W_z\bar{\mathbf{k}}$

The commutative property of vector addition allows the terms to be rearranged as: $\bar{\mathbf{U}} + \bar{\mathbf{V}} + \bar{\mathbf{W}} = \bar{\mathbf{R}} = (U_x + V_x + W_x)\bar{\mathbf{i}} +$ (____________)$\bar{\mathbf{j}}$ + (____________) $\bar{\mathbf{k}}$.

$U_y + V_y + W_y$
$U_z + V_z + W_z$

With rectangular components and unit vectors, we have reduced *vector* addition to addition of _____ (type) components.

scalar

We now investigate techniques of transforming and changing the description of vectors. Such techniques are needed when vector quantities are specified in forms inconvenient for computational purposes. One useful representation is a magnitude and a unit vector giving direction. That is, a vector $\bar{\mathbf{F}}$ is expressed by use of its magnitude F and the unit vector $\bar{\mathbf{e}}_f$ parallel to $\bar{\mathbf{F}}$. In equation form we write this as _____.

$\bar{\mathbf{F}} = F\bar{\mathbf{e}}_f$

Thus the unit vector $\bar{\mathbf{e}}_f$ equals $\bar{\mathbf{F}}$ divided by _________________.

F, the magnitude of $\bar{\mathbf{F}}$

Now we will establish this type of representation, given a vector in somewhat different form. Suppose we begin with a vector $\bar{\mathbf{F}}$ as $F_x\bar{\mathbf{i}} + F_y\bar{\mathbf{j}} + F_z\bar{\mathbf{k}}$. Since $\bar{\mathbf{i}}$, $\bar{\mathbf{j}}$, and $\bar{\mathbf{k}}$ are orthogonal, the magnitude F of the vector equals the square root of the sum, $F_x^2 +$ __ $+$ __.

F_y^2 F_z^2

The unit vector $\bar{\mathbf{e}}_f$ is found by dividing $\bar{\mathbf{F}}$ by its scalar magnitude. If $\bar{\mathbf{F}}$ is expressed in component form as $\bar{\mathbf{F}} = F_x\bar{\mathbf{i}} + F_y\bar{\mathbf{j}} + F_z\bar{\mathbf{k}}$, then $\bar{\mathbf{e}}_f = \bar{\mathbf{F}}/F = (F_x/F)\bar{\mathbf{i}} +$ (___)$\bar{\mathbf{j}}$ + _____.

F_y/F
$(F_z/F)\bar{\mathbf{k}}$

Thus the scalar components of the unit vector $\bar{\mathbf{e}}_f$ are e_{fx} = ___, e_{fy} = ___, e_{fz} = ___.

F_x/F F_y/F F_z/F

We now consider an example, the force $\overline{\mathbf{F}} = 3\bar{\mathbf{i}} + 4\bar{\mathbf{j}} + 12\bar{\mathbf{k}}$ lb. On scratch paper, compute the magnitude F as equal to __ lb.

13

Next, write the unit vector $\overline{\mathbf{e}_f}$ as $3\bar{\mathbf{i}}/13 + 4\bar{\mathbf{j}}/$__ + (___) $\bar{\mathbf{k}}$.

13

12/13

A quick check of this result shows that the magnitude of the unit vector $\bar{\mathbf{e}}_f$ {equal to the square root of $\left(\frac{3}{13}\right)^2 + \left(\frac{4}{13}\right)^2 + \left(\frac{12}{13}\right)^2$} equals ___.

one

Thus the unit vector representation, $\overline{\mathbf{F}} = F\overline{\mathbf{e}}$, is $\overline{\mathbf{F}} = 13$ (____________________).

$(3/13)\bar{\mathbf{i}} + (4/13)\bar{\mathbf{j}} + (12/13)\bar{\mathbf{k}} = \bar{\mathbf{e}}_f$

We just changed a component representation to a representation using a magnitude and a unit vector. Let's reverse this procedure and find the components of $\overline{\mathbf{F}}$ given the unit vector representation. Simply multiply the components of the unit vector by the magnitude of $\overline{\mathbf{F}}$ to find the magnitude of each ________ of $\overline{\mathbf{F}}$.

component

Another common vector specification is the magnitude of the vector plus two points on its line of action. Figure 2.9 shows a vector $\overline{\mathbf{P}}$, magnitude 20 lb,

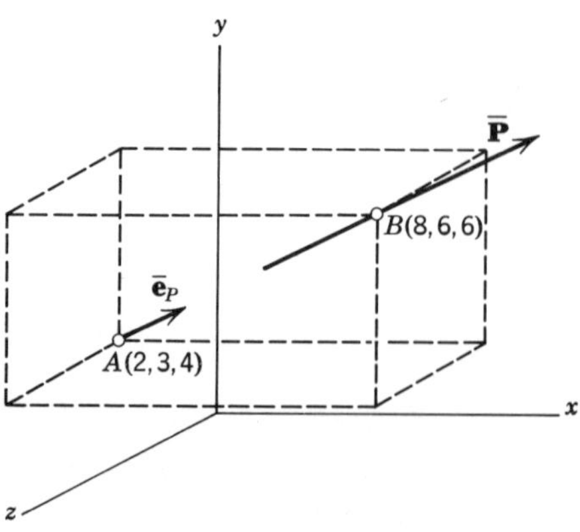

Figure 2.9

with a line of action through points A and B, positive from A to B. Here we know the magnitude of the vector and we seek a way to express its ____________.

orientation/direction

We thus need a unit vector along the line of action

of $\overline{\mathbf{P}}$, positive from A to B. For this purpose we temporarily *ignore* $\overline{\mathbf{P}}$ and use geometry to find this unit vector. Consider the rectangular parallelepiped shown dotted in Fig. 2.9 and in Fig. 2.10. The auxiliary vector

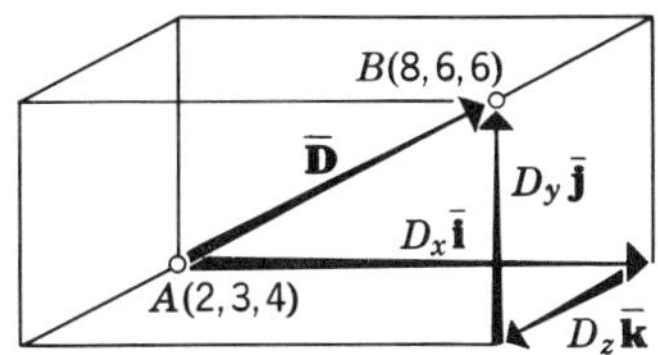

Figure 2.10

$\overline{\mathbf{D}}$ from point A to point B has components D_x, D_y, and D_z. These components are given by differences in the coordinates of A and B. For example, D_x is the difference between the x coordinate of B, (8), and the x coordinate of A, (2). That is, D_x equals __. 6

Likewise, $D_y = y_B - y_A = 6 -$ __ $=$ __. Finally, 3 3
$D_z =$ __ $-$ __ $=$ __. 6 4 2

We have $D_x = 6$, $D_y = 3$, and $D_z = 2$. Therefore, the magnitude of $\overline{\mathbf{D}} =$ __. 7

We can write the unit vector $\bar{\mathbf{e}}_f$, directed from A to B, using the relation $\bar{\mathbf{e}}_p = (D_x/D)\,\bar{\mathbf{i}} + (D_y/D)\,\bar{\mathbf{j}} + (D_z/D)\,\bar{\mathbf{k}}$, as $\bar{\mathbf{e}}_p =$ ___$\bar{\mathbf{i}}$ + ___$\bar{\mathbf{j}}$ + ___$\bar{\mathbf{k}}$. 6/7 3/7 2/7

Now, recall the vector $\overline{\mathbf{P}}$ with magnitude $P = 20$ lb. Since $\bar{\mathbf{e}}_p$ is parallel to $\overline{\mathbf{P}}$, we write $\overline{\mathbf{P}}$ as $\overline{\mathbf{P}} = P\bar{\mathbf{e}}_p$. Inserting appropriate numerical values, $\overline{\mathbf{P}} =$ __ (___$\bar{\mathbf{i}}$ + ___$\bar{\mathbf{j}}$ + ___$\bar{\mathbf{k}}$). 20 6/7 3/7 2/7

Remember that in determining the unit vector $\bar{\mathbf{e}}_p$, you ________ use the magnitude of the vector $\overline{\mathbf{P}}$ and do not
do/do not
________ use the location of points A and B. do
do/do not

With the magnitude $P = 20$ lb and the unit vector $\bar{\mathbf{e}}_p$, you can readily determine any other characteristic of $\overline{\mathbf{P}}$. For example, $P_x =$ ____, $P_y =$ ____, and P_z $=$ ____. 120/7 60/7 40/7

Next we review a useful vector operation, the scalar (or dot) product of two vectors. Consider the vectors $\overline{\mathbf{P}}$ and $\overline{\mathbf{Q}}$ in Fig. 2.11. Their scalar product is defined as $\overline{\mathbf{P}}\cdot\overline{\mathbf{Q}} = PQ\cos\alpha$ where α is the angle between the *positive* directions of the two vectors. The

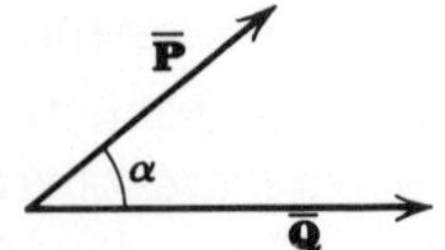

Figure 2.11

90° $\left(\frac{\pi}{2}\text{ rad}\right)$

scalar product is positive or negative, depending upon whether α is greater or less than ________.

The order in this product is unimportant, ($\overline{\mathbf{A}}\cdot\overline{\mathbf{B}} = \overline{\mathbf{B}}\cdot\overline{\mathbf{A}}$). The scalar product is also distributive, i.e., $\overline{\mathbf{A}}\cdot(\overline{\mathbf{B}}+\overline{\mathbf{C}}) = \overline{\mathbf{A}}\cdot\overline{\mathbf{B}} + \overline{\mathbf{A}}\cdot\overline{\mathbf{C}}$.

The scalar product can be used to find components of vectors and angles between vectors. For example, let's determine the component of $\overline{\mathbf{P}}$ in the $\overline{\mathbf{Q}}$ direction where $\overline{\mathbf{P}}$ and $\overline{\mathbf{Q}}$ are arbitrary vectors in space. Figure 2.12 shows that we must find the length of the

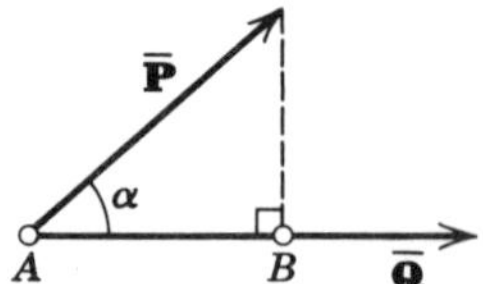

Figure 2.12

line segment AB. From geometry the length of AB equals P cos __.

α

However, the scalar product $\overline{\mathbf{P}}\cdot\overline{\mathbf{Q}}$ equals $PQ \cos\alpha$. Therefore, the component of $\overline{\mathbf{P}}$ in the $\overline{\mathbf{Q}}$ direction equals $\overline{\mathbf{P}}\cdot\overline{\mathbf{Q}}$ divided by __.

Q

$\overline{\mathbf{Q}}$ divided by Q is the unit vector $\bar{\mathbf{e}}_q$. Thus the projection of $\overline{\mathbf{P}}$ in the $\overline{\mathbf{Q}}$ direction is the scalar product of of $\overline{\mathbf{P}}$ with $\bar{\mathbf{e}}_q$. The projection of $\overline{\mathbf{P}}$ on $\overline{\mathbf{Q}}$ is written as $\overline{\mathbf{P}}\cdot$ __.

$\bar{\mathbf{e}}_q$

To find the *angle* between the lines defined by the vectors $\overline{\mathbf{P}}$ and $\overline{\mathbf{Q}}$, write $\overline{\mathbf{P}}\cdot\overline{\mathbf{Q}} = PQ\cos\alpha$ and solve for $\cos\alpha$ by dividing $\overline{\mathbf{P}}\cdot\overline{\mathbf{Q}}$ by the product __.

PQ

This is equivalent to taking the scalar product of two unit vectors, one parallel to $\overline{\mathbf{P}}$ and the other to $\overline{\mathbf{Q}}$. Thus, $\cos\alpha$ = (__) · (__).

$\bar{\mathbf{e}}_p$ $\quad$ $\bar{\mathbf{e}}_q$

We have shown that it is simple to find orthogonal projections and angles between lines using scalar products. To calculate scalar products we shall use vectors $\overline{\mathbf{P}}$ and $\overline{\mathbf{Q}}$ in orthogonal component form. Begin by writing $\overline{\mathbf{P}} = P_x\overline{\mathbf{i}} + P_y\overline{\mathbf{j}} + P_z\overline{\mathbf{k}}$ and $\overline{\mathbf{Q}} = Q_x\overline{\mathbf{i}} + Q_y\overline{\mathbf{j}} + Q_z\overline{\mathbf{k}}$. The scalar product, $\overline{\mathbf{P}}\cdot\overline{\mathbf{Q}}$, is written as $(P_x\overline{\mathbf{i}} + P_y\overline{\mathbf{j}} + P_z\overline{\mathbf{k}})\cdot(Q_x\overline{\mathbf{i}} + Q_y\overline{\mathbf{j}} + Q_z\overline{\mathbf{k}})$. Expanding, using the distributive property, yields nine terms like $P_x\overline{\mathbf{i}}\cdot Q_x\overline{\mathbf{i}}$ and $P_y\overline{\mathbf{j}}\cdot Q_z\overline{\mathbf{k}}$. Consider the product $P_x\overline{\mathbf{i}}\cdot Q_x\overline{\mathbf{i}}$, rewriting it as $P_xQ_x\overline{\mathbf{i}}\cdot\overline{\mathbf{i}}$. From the definition, the scalar product of the unit vector $\overline{\mathbf{i}}$ with itself equals the square of the magnitude of $\overline{\mathbf{i}}$ and the cosine of the included angle (zero degrees). Thus, $\overline{\mathbf{i}}\cdot\overline{\mathbf{i}}$ equals ___. one

Similarly, $\overline{\mathbf{j}}\cdot\overline{\mathbf{j}} = \overline{\mathbf{k}}\cdot\overline{\mathbf{k}} =$ ___. one

Contrast, however, the scalar product of $P_y\overline{\mathbf{j}}\cdot Q_z\overline{\mathbf{k}}$. The vectors $\overline{\mathbf{j}}$ and $\overline{\mathbf{k}}$ are not parallel but are ______. perpendicular/orthogonal

Since they are perpendicular (or orthogonal), the cosine of the angle between them is ___, and their scalar product equals ___. zero zero

Since $\overline{\mathbf{i}}$, $\overline{\mathbf{j}}$, and $\overline{\mathbf{k}}$ are mutually orthogonal (perpendicular), any scalar product involving two different unit vectors of the set $\overline{\mathbf{i}}$, $\overline{\mathbf{j}}$, and $\overline{\mathbf{k}}$ will equal ___. zero

We conclude that, of the nine terms of our expansion, three involve the products $\overline{\mathbf{i}}\cdot\overline{\mathbf{i}} = \overline{\mathbf{j}}\cdot\overline{\mathbf{j}} = \overline{\mathbf{k}}\cdot\overline{\mathbf{k}} = 1$ and six involve products of the form $\overline{\mathbf{i}}\cdot\overline{\mathbf{j}}$ and $\overline{\mathbf{k}}\cdot\overline{\mathbf{j}}$ that equal ___. zero

Thus, with $\overline{\mathbf{P}}$ and $\overline{\mathbf{Q}}$ in component form $\overline{\mathbf{P}}\cdot\overline{\mathbf{Q}}$ reduces from nine to three terms. These three nonzero terms are associated with the scalar products of the two $\overline{\mathbf{i}}$, the two $\overline{\mathbf{j}}$, and the two $\overline{\mathbf{k}}$ components. Therefore, $\overline{\mathbf{P}}\cdot\overline{\mathbf{Q}} = P_xQ_x + P_yQ_y +$ ___. P_zQ_z

We will now use the scalar product to determine the projection of $\overline{\mathbf{Q}}$ on $\overline{\mathbf{P}}$ in Fig. 2.13. $\overline{\mathbf{P}}$ is specified by its end points *A* and *B*. $\overline{\mathbf{Q}}$ is given in component form and passes through *C*. First we express $\overline{\mathbf{P}}$ in component form as $\overline{\mathbf{P}} = -3\overline{\mathbf{i}} +$ __$\overline{\mathbf{j}} +$ __$\overline{\mathbf{k}}$. 4 0

The projection of $\overline{\mathbf{Q}}$ on $\overline{\mathbf{P}}$ equals $\overline{\mathbf{Q}}\cdot\overline{\mathbf{P}}$ divided by the magnitude of __. The magnitude of $\overline{\mathbf{P}}$ equals __. $\overline{\mathbf{P}}$ 5

Thus the projection of $\overline{\mathbf{Q}}$ on $\overline{\mathbf{P}}$ equals $(Q_xP_x + Q_yP_y + Q_zP_z)$ divided by __. 5

Perform the calculations to find that the projection of $\overline{\mathbf{Q}}$ on $\overline{\mathbf{P}}$ equals ___. 14/5

Finally, let's find the angle between these vectors.

This angle is the angle formed when we translate either vector so that it intersects the other. To find

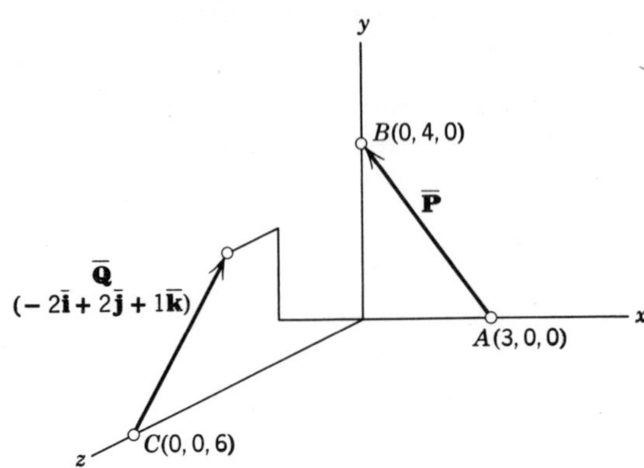

Figure 2.13

the angle, we need the magnitudes of $\overline{\mathbf{P}}$ and $\overline{\mathbf{Q}}$. P is
3 known; determine $Q \cdot Q$ equals __.
PQ Now use the formula $\cos \alpha = \overline{\mathbf{P}} \cdot \overline{\mathbf{Q}}$ divided by __.
You have already found that $\overline{\mathbf{P}} \cdot \overline{\mathbf{Q}} = (Q_x P_x + Q_y P_y$
14 $+ Q_z P_z)$ equals __.
14/15 Since $P = 5$ and $Q = 3$, $\cos \alpha$ equals ____.
Using a slide rule or trig table, find that α equals
21 __ degrees.

Summary

Whereas scalars are fully characterized by their magnitude, vector quantities involve a magnitude, an orientation, and a special rule of addition—the parallelogram rule. The most useful representation of vectors is in terms of rectangular components and a set of orthogonal unit vectors, denoted here by $\overline{\mathbf{i}}$, $\overline{\mathbf{j}}$, and $\overline{\mathbf{k}}$. Vectors are also specified using a magnitude and the angles formed with the coordinate directions or two points on the line of action. Changes from one representation to another are readily made using direction cosines and the fact that the square of the diagonal of a rectangular parallelepiped equals the sum of the squares of the sides. Finally, the scalar product of two vectors $\overline{\mathbf{P}}$ and $\overline{\mathbf{Q}}$, defined as $\overline{\mathbf{P}} \cdot \overline{\mathbf{Q}} = PQ \cos \alpha$, is expressed in rectangular component form as $\overline{\mathbf{P}} \cdot \overline{\mathbf{Q}} = P_x Q_x + P_y Q_y + P_z Q_z$ and is used to determine angles between lines and projections.

Problems

(1) Classify the following quantities as scalars, vectors, or neither: (a) force, (b) temperature, (c) velocity, (d) displacement, (e) finite rotation, (f) time, (g) the array of coefficients of a set of three linear algebraic equations in three unknowns.

(2) Find the sum $\overline{\mathbf{R}}$ of the vectors $\overline{\mathbf{A}}$, $\overline{\mathbf{B}}$, and $\overline{\mathbf{C}}$ in Fig. P2.1 by graphical techniques.

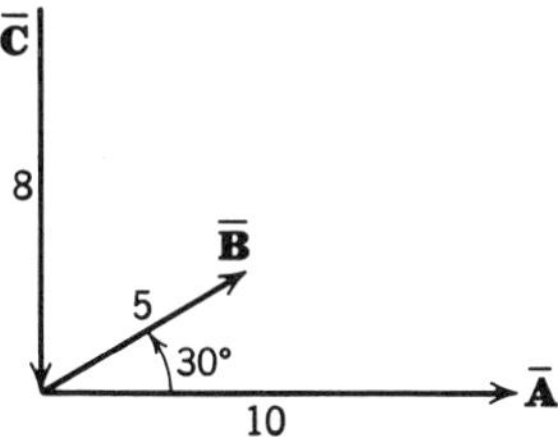

Figure P2.1

(3) Find the sum $\overline{\mathbf{R}}$ of the vectors $\overline{\mathbf{A}}$, $\overline{\mathbf{B}}$, and $\overline{\mathbf{C}}$ in Fig. P2.1 by trigonometric techniques.

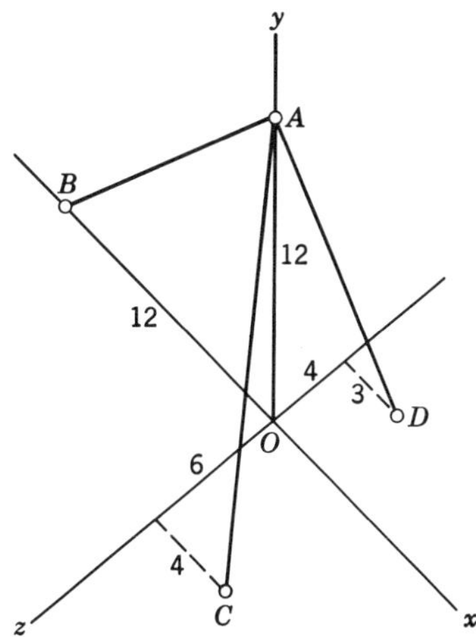

Figure P2.2

(4) A force $\overline{\mathbf{F}}$ is given in rectangular components as

$$\overline{\mathbf{F}} = \overline{\mathbf{i}} - 2\overline{\mathbf{j}} + 2\overline{\mathbf{k}}$$

What is its magnitude? What is the unit vector $\overline{\boldsymbol{\lambda}}$ in the direction of $\bar{\mathbf{F}}$? Does it make sense to write a vector in rectangular components without defining reference axes?

(5) Find $\bar{\mathbf{A}} \cdot \bar{\mathbf{B}}$ where

$$\bar{\mathbf{A}} = -\bar{\mathbf{i}} - 2\bar{\mathbf{j}} + 2\bar{\mathbf{k}} \qquad \bar{\mathbf{B}} = 4\bar{\mathbf{i}} + 4\bar{\mathbf{j}} + 2\bar{\mathbf{k}}$$

What is the angle between $\bar{\mathbf{A}}$ and $\bar{\mathbf{B}}$? What is the projection of $\bar{\mathbf{A}}$ on $\bar{\mathbf{B}}$? What is the projection of $\bar{\mathbf{B}}$ on $\bar{\mathbf{A}}$?

(6) A pole *OA* in Fig. P2.2 is restrained by three wires *AB*, *AC*, and *AD*. In order to analyze the forces in the cables we must make use of the fact that the cables support loads only in their own direction. The directions of the cable forces are thus known and only their magnitudes need be determined.

What is the vector $\overline{\mathbf{AB}}$ from *A* to *B*? What is its magnitude? What is the unit vector $\bar{\lambda}_{AB}$ parallel to $\overline{\mathbf{AB}}$? Show that the force $\bar{\mathbf{F}}_{AB}$ in cable *AB* must then be $\bar{\mathbf{F}}_{AB} = F_{AB}\,\bar{\lambda}_{AB}$. What are the unit vectors $\bar{\lambda}_{AC}$ and $\bar{\lambda}_{AD}$? Express the forces in cables *AC* and *AD* in terms of their respective magnitudes and unit vectors.

Answers to Problems

Chapter 2

(1) *a*, *c*, *d* vectors
b, *f* scalars
e, *g* neither

(2) and (3) $\bar{\mathbf{R}} = 14.33\bar{\mathbf{i}} - 5.5\bar{\mathbf{j}}$

(4) $F = 3$

$$\bar{\lambda}_F = \frac{1}{3}(\bar{\mathbf{i}} - 2\bar{\mathbf{j}} + 2\bar{\mathbf{k}})$$

no

(5) $\bar{\mathbf{A}} \cdot \bar{\mathbf{B}} = -8$

$\theta = 116.4°$

$A\cos\theta = -1.322$

$B\cos\theta = -2.644$

(6) $$\overline{\mathbf{AB}} = AB\,\bar{\lambda}_{AB} = 12\sqrt{2}\left[-\frac{1}{\sqrt{2}}(\bar{\mathbf{i}} + \bar{\mathbf{j}})\right]$$

$$\bar{\mathbf{F}}_{AB} = \frac{-F_{AB}}{\sqrt{2}}(\bar{\mathbf{i}} + \bar{\mathbf{j}})$$

$$\bar{\mathbf{F}}_{AC} = \frac{F_{AC}}{14}(4\bar{\mathbf{i}} - 12\bar{\mathbf{j}} + 6\bar{\mathbf{k}})$$

$$\bar{\mathbf{F}}_{AD} = \frac{F_{AD}}{13}(3\bar{\mathbf{i}} - 12\bar{\mathbf{j}} - 4\bar{\mathbf{k}})$$

chapter 3

Moments of Forces

Objectives

In this chapter we examine the torque or twisting effects produced by a force. The principal tool used is the vector product of two vectors. However, since the scalar component approach is frequently simpler, it is presented also. Successful completion of this chapter should make it possible for the student to:

1. Explain the concepts of the moment of a force about a point and about a line.
2. Define the vector product of two vectors, list its properties, and correctly calculate vector products using the orthogonal component description of two vectors.
3. Define the moment of a force about a point using the vector product, and calculate moments using this general technique and the scalar force component approach.
4. Define the moment of a force about a line in terms of the scalar triple product and determine such moments using the general vector product definition or using properly chosen scalar force components.
5. Examine typical statics problems requiring moments about points and lines and determine whether the general vector approach or scalar component approach is more suitable.
6. Define a couple, state its characteristics, and correctly identify couples that are equivalent.

A general characteristic of forces is that they tend to turn or rotate the bodies on which they act, as well as to make these bodies translate (i.e., move without rotation). In this chapter we study the measure of this effect, called the moment of a force. In other words, we shall learn to find and use the *moment of a force*, which is a measure of the tendency of a force to produce rotation. The word moment is here synonymous with a torque or twisting effect.

When we consider moments, we encounter a property or effect of forces that makes them more than just vectors. In particular, not only must we designate the magnitude and direction of a force but, to determine its moment effect, we must also designate a specific line along which the force acts. This we call the *line of action* of the force. Thus, where moments are concerned, we must know the magnitude and direction of a force and also its ________.

line of action

We shall first review the second type of vector multiplication, the vector (or cross) product of two vectors. Figure 3.1 shows two vectors, $\bar{\mathbf{A}}$ and $\bar{\mathbf{B}}$, with

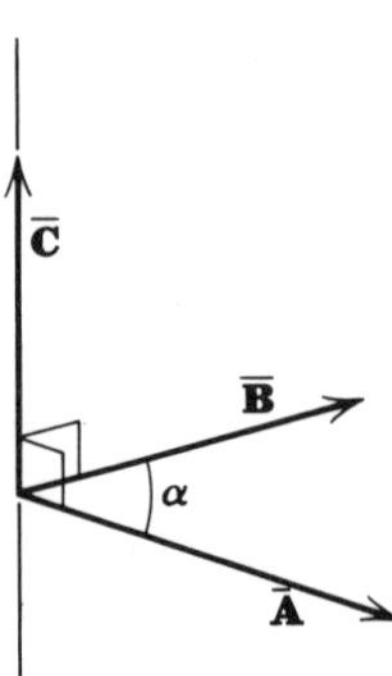

Figure 3.1

the angle α between them. We define the vector product $\bar{\mathbf{C}} = \bar{\mathbf{A}} \times \bar{\mathbf{B}}$ as *a vector*, of magnitude $AB \sin \alpha$, directed perpendicular to the plane defined by the vectors $\bar{\mathbf{A}}$ and $\bar{\mathbf{B}}$. The vector $\bar{\mathbf{C}}$ has a positive direction, in accord with the right-hand rule, for the rotation of the vector $\bar{\mathbf{A}}$ *into the vector* $\bar{\mathbf{B}}$. The positive direction is the direction that a right-hand screw would advance along the $\bar{\mathbf{C}}$ axis if rotated through α from $\bar{\mathbf{A}}$ into $\bar{\mathbf{B}}$. Let us now observe some of the general properties of the cross product that are implied by the definition. First, what is the relationship between the cross product $\bar{\mathbf{A}} \times \bar{\mathbf{B}}$ and the product $\bar{\mathbf{B}} \times \bar{\mathbf{A}}$? Since

the magnitude of $\overline{\mathbf{A}} \times \overline{\mathbf{B}}$ equals $AB \sin \alpha$ and the magnitude $\overline{\mathbf{B}} \times \overline{\mathbf{A}}$ equals $BA \sin \alpha$, the magnitude of the products will be ________________ in the two cases. identical
identical/different

What will be the direction of the vector $\overline{\mathbf{C}}$? For the product $\overline{\mathbf{A}} \times \overline{\mathbf{B}}$, we have rotated $\overline{\mathbf{A}}$ through α into $\overline{\mathbf{B}}$ to obtain the positive direction of $\overline{\mathbf{C}}$ as shown in Fig. 3.1. The cross product $\overline{\mathbf{B}} \times \overline{\mathbf{A}}$ would involve a rotation of the vector $\overline{\mathbf{B}}$ through α into the vector $\overline{\mathbf{A}}$ and thus would have the ______________ direction as $\overline{\mathbf{C}}$. We opposite
same/opposite
conclude that $\overline{\mathbf{A}} \times \overline{\mathbf{B}}$ equals ___________ $\overline{\mathbf{B}} \times \overline{\mathbf{A}}$. minus
plus/minus

Another important property of the cross product is that it is distributive. That is $\overline{\mathbf{A}} \times (\overline{\mathbf{D}} + \overline{\mathbf{E}}) = \overline{\mathbf{A}} \times \overline{\mathbf{D}} + \overline{\mathbf{A}} \times \overline{\mathbf{E}}$.

Since $\overline{\mathbf{A}} \times \overline{\mathbf{B}}$ does not equal $\overline{\mathbf{B}} \times \overline{\mathbf{A}}$ but rather is equal to _____ $\overline{\mathbf{B}} \times \overline{\mathbf{A}}$, we must preserve the order in which terms are presented in cross products. Thus we conclude that $\overline{\mathbf{A}} \times (\overline{\mathbf{D}} + \overline{\mathbf{E}})$ ________ $= \overline{\mathbf{A}} \times \overline{\mathbf{D}} + \overline{\mathbf{E}} \times \overline{\mathbf{A}}$. minus / is not
is/is not

Now that we have defined and examined some properties of the vector product of two vectors, let us consider a way of computing this product. Here, as with the scalar product, we first express our vectors in component form. Thus we write $\overline{\mathbf{A}} = A_x\overline{\mathbf{i}} + A_y\overline{\mathbf{j}} + A_z\overline{\mathbf{k}}$ and $\overline{\mathbf{B}} =$ ______________. $B_x\overline{\mathbf{i}} + B_y\overline{\mathbf{j}} + B_z\overline{\mathbf{k}}$

We now perform the vector product operation with $\overline{\mathbf{A}}$ and $\overline{\mathbf{B}}$ in their component forms. First, using the distributive rule, we note that $(A_x\overline{\mathbf{i}} + A_y\overline{\mathbf{j}} + A_z\overline{\mathbf{k}}) \times (B_x\overline{\mathbf{i}} + B_y\overline{\mathbf{j}} + B_z\overline{\mathbf{k}})$ will yield nine terms of the form $(A_x\overline{\mathbf{i}}) \times (B_x\overline{\mathbf{i}})$, $(A_y\overline{\mathbf{j}}) \times (B_z\overline{\mathbf{k}})$, and so forth. To evaluate these terms, we examine the cross product behavior of the unit vector combinations. First, the cross product of two parallel vectors, that is, $\overline{\mathbf{i}} \times \overline{\mathbf{i}}$ or $\overline{\mathbf{j}} \times \overline{\mathbf{j}}$, will be equal to ____, since the sine of zero degrees is ____. zero / zero

Furthermore, $\overline{\mathbf{i}}$, $\overline{\mathbf{j}}$, and $\overline{\mathbf{k}}$ are mutually perpendicular; thus the angle between any of these vectors is 90 degrees. Since they are defined as a right-handed coordinate system, we know that $\overline{\mathbf{i}} \times \overline{\mathbf{j}}$ must be in the $\overline{\mathbf{k}}$ direction, $\overline{\mathbf{j}} \times \overline{\mathbf{k}}$ must be in the $\overline{\mathbf{i}}$ direction, and $\overline{\mathbf{k}} \times \overline{\mathbf{i}}$ must be in the __ direction. $\overline{\mathbf{j}}$

Reversing the order of the terms in a vector product changes the sign of the resultant, so $\bar{j} \times \bar{i} = -\bar{k}$, $\bar{k} \times \bar{j} = -$__ and __ $\times$ __ $= -\bar{j}$.

$\bar{i}$ $\bar{i}$ $\bar{k}$

From these properties of $\bar{i}$, $\bar{j}$, and $\bar{k}$ we conclude that of the nine terms of the vector product of $\bar{A}$ and $\bar{B}$ in component form, three are zero, three will have positive signs, and three will have ______ signs.

negative

Now, perform this operation term by term. If properly done, you will find that $\bar{A} \times \bar{B} = (A_yB_z - A_zB_y)\bar{i} + ($________$)\bar{j} + ($________$)\bar{k}$.

$A_zB_x - A_xB_z$ $A_xB_y - A_yB_x$

There is an easy way to remember this that will save time and reduce mistakes. The following determinant, when expanded, yields the desired product $\bar{A} \times \bar{B}$.

$$\bar{A} \times \bar{B} = \begin{vmatrix} \bar{i} & \bar{i} & \bar{k} \\ A_x & A_y & A_z \\ B_x & B_y & B_z \end{vmatrix}$$

In the space below, write the expansion of this determinant (by minors of the first row) and demonstrate that you obtain the results given above.

We now introduce the concept of the moment of a force about a point. We are concerned with the twisting effect that a force has about a point in space. This twisting effect, or moment, involves a *magnitude* dependent upon the magnitude of the force and the perpendicular distance from the line of action of this force to the point. The way this force tends to rotate the point, or the body associated with the point, depends upon the orientation of the force relative to the point. Thus the moment of a force about a point has a magnitude and an orientation and is therefore a ____ quantity.

vector

We now examine the moment of a force $\bar{F}$ about the point O in space. The line of action of the force $\bar{F}$

and the point O define a plane in space, as in Fig. 3.2. We choose our coordinate system so that the x-y plane

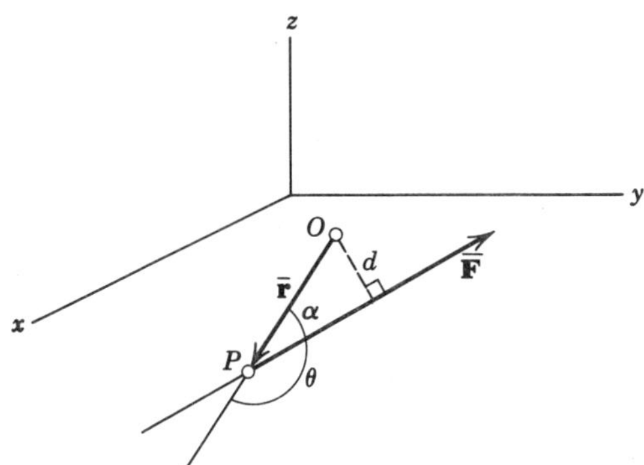

Figure 3.2

is this plane defined by $\bar{\mathbf{F}}$ and O. To define the moment of the force $\bar{\mathbf{F}}$ about the point O, we first choose a point P on the line of action of the force $\bar{\mathbf{F}}$. A vector $\bar{\mathbf{r}}$ is then defined as the vector *from O to P*. The moment $\bar{\mathbf{M}}$ of the force $\bar{\mathbf{F}}$ about point O is defined as $\bar{\mathbf{M}} = \bar{\mathbf{r}} \times \bar{\mathbf{F}}$. With this definition we see that the moment $\bar{\mathbf{M}}$ is defined as a ____________ quantity. The direction of   vector
vector/scalar
moment vector, $\bar{\mathbf{M}}$, will be parallel to the __ axis, since   z
by definition $\bar{\mathbf{r}} \times \bar{\mathbf{F}}$ is perpendicular to the x-y plane.

The definition of the cross product also shows that the magnitude of the moment will be proportional to the magnitude of the force $\bar{\mathbf{F}}$ and the shortest distance, d, between point O and the line of action of $\bar{\mathbf{F}}$. This follows from the geometry of Fig. 3.2. From the definition, $\bar{\mathbf{M}} = \bar{\mathbf{r}} \times \bar{\mathbf{F}}$. We have $\bar{\mathbf{M}} = F(r \sin \theta)\bar{\mathbf{n}}$ where $\bar{\mathbf{n}}$ is a unit vector normal to the plane defined by $\bar{\mathbf{F}}$ and O.

Since $\sin \theta$ equals $\sin \alpha$, $r \sin \theta$ equals $r \sin \alpha$
$=$ __. d

The expression for $\bar{\mathbf{M}}$ can be written $\bar{\mathbf{M}} = Fd\bar{\mathbf{n}}$. The definition of the cross product assures that regardless of the point chosen for P on the line of action of $\bar{\mathbf{F}}$, the product expression will involve the shortest distance from O to $\bar{\mathbf{F}}$ and thus we may choose ___ point   any
on the line of action of $\bar{\mathbf{F}}$ for this operation.

We shall try to choose a point on the line of action

of $\bar{\mathbf{F}}$ for which the computations are easy. For example, if the force $\bar{\mathbf{F}}$ were defined in terms of two points on its line of action, we might choose either of these points.

In summary, the moment of a force about a point is represented by a vector with a magnitude proportional to the product of the force and the shortest distance between the line of action of the force and the point. The rotational or twisting effect defined by this vector acts along an axis perpendicular to the plane containing the force and the point. The positive
right-hand direction of this rotation is given by the ______ rule used in the definition of the cross product.

A popular method of determining the moment of a force about a point rests upon taking components of the force in a certain way. In Fig. 3.3 the *x-y* plane

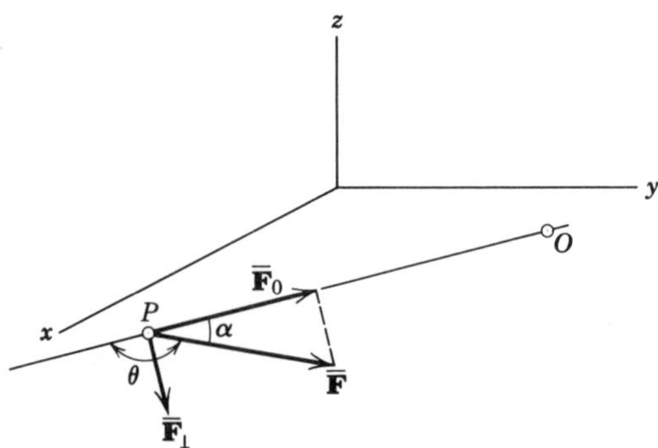

Figure 3.3

is the plane defined by $\bar{\mathbf{F}}$ and O. The common practice is to take components of the force $\bar{\mathbf{F}}$ so that the line of action of one ($\bar{\mathbf{F}}_o$) passes through the point O. The other component ($\bar{\mathbf{F}}_\perp$) is taken perpendicular to $\bar{\mathbf{F}}_o$. We will use the principle of moments which states that the sum of the individual moments of the components of a force is equal to the moment of the force itself. Examine the moment of $\bar{\mathbf{F}}_o$ about O. The line of action of $\bar{\mathbf{F}}_o$ passes through the point O. The moment con-
zero tribution of $\bar{\mathbf{F}}_o$ equals ____ because the perpendicular
zero distance from O to the line of action of $\bar{\mathbf{F}}_o$ is ____.

Take $\bar{\mathbf{r}}$ as the vector from O to P. Then $\bar{\mathbf{r}} \times \bar{\mathbf{F}}_\perp$ will equal $rF_\perp\bar{\mathbf{n}}$, since $\bar{\mathbf{F}}_\perp$ is perpendicular to $\bar{\mathbf{r}}$. Notice that this moment equals the moment obtained previously when we consider $F_\perp$ as F sin __. Hence, in both cases the moment, $\bar{\mathbf{M}}$, of $\bar{\mathbf{F}}$ about O equals Fr sin __$\bar{\mathbf{n}}$. α — θ or α ($\sin\theta = \sin\alpha$)

We have shown that moments may be calculated either directly by application of the cross product definition or by splitting the force into components and summing the moment contributions of each component. Through practice, you will learn to choose the most efficient method for a particular problem.

As an example, we determine the moment of the force $\bar{\mathbf{F}}$ in Fig. 3.4 about point O. $\bar{\mathbf{F}}$ has a magnitude of

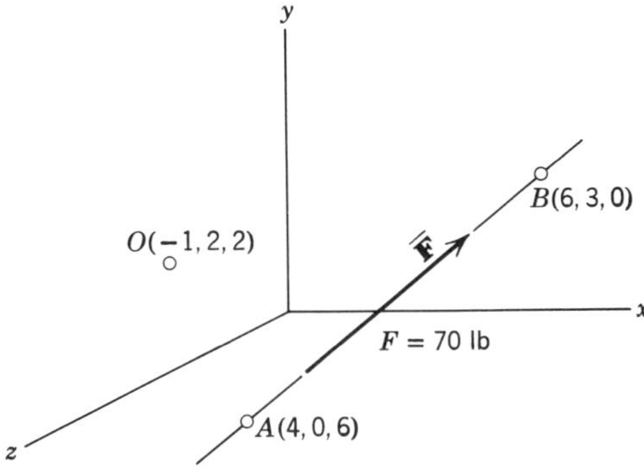

Figure 3.4

70 lb, and its line of action is from A to B. Since the moment definition requires a cross product operation, we first express $\bar{\mathbf{F}}$ in vector form and choose a vector $\bar{\mathbf{r}}$ from O to a point on the line of action of $\bar{\mathbf{F}}$. The vector form of $\bar{\mathbf{F}}$ used will be a magnitude and a direction given by a unit vector from A to B, $\bar{\mathbf{e}}_F$. To find $\bar{\mathbf{e}}_F$ we need the vector $\overline{\mathbf{AB}}$ from A to B. This vector has components $(AB)_x$ = __, $(AB)_y$ = __, and $(AB)_z$ = __. 2 3 −6

The length of this vector $\overline{\mathbf{AB}}$ is __________. $\sqrt{6^2 + 3^2 + 2^2} = 7$

Thus $\bar{\mathbf{e}}_F = 2/7\bar{\mathbf{i}}$ + ___$\bar{\mathbf{j}}$ + ___$\bar{\mathbf{k}}$. 3/7 −6/7

In our notation, the vector $\bar{\mathbf{F}}$ is $\bar{\mathbf{F}}$ = 70 __ lb. $\bar{\mathbf{e}}_F$

In component from, $\bar{\mathbf{F}}$ is written as $\bar{\mathbf{F}}$ = __$\bar{\mathbf{i}}$ + __$\bar{\mathbf{j}}$ 20 30
+ ___$\bar{\mathbf{k}}$ lb. −60

We now choose a vector $\bar{\mathbf{r}}$ from O to some point P on the line of action of $\bar{\mathbf{F}}$. Since the locations of two points on the line of action of $\bar{\mathbf{F}}$ are known, it is convenient to choose for the point P, either the point A or B.

We take our vector $\bar{\mathbf{r}}$ from O to point B. Then the

7 1 −2

vector $\bar{\mathbf{r}}$ is written as $\bar{\mathbf{r}} =$ __$\bar{\mathbf{i}}$ + __$\bar{\mathbf{j}}$ + __$\bar{\mathbf{k}}$.

$$\begin{vmatrix} \bar{\mathbf{i}} & \bar{\mathbf{j}} & \bar{\mathbf{k}} \\ 7 & 1 & -2 \\ 20 & 30 & -60 \end{vmatrix}$$

The moment $\bar{\mathbf{M}}$ about O is $\bar{\mathbf{M}}_O = \bar{\mathbf{r}} \times \bar{\mathbf{F}}$. Express $\bar{\mathbf{M}}_O$ in determinant form (on scratch paper).

Expansion of this determinant by elements of the

0 380 190

first row shows that $\bar{\mathbf{M}}_O =$ __$\bar{\mathbf{i}}$ + __$\bar{\mathbf{j}}$ + __$\bar{\mathbf{k}}$.

Since any point on the line of action of $\bar{\mathbf{F}}$ can be chosen for the vector $\bar{\mathbf{r}}$ let us recalculate the moment

5

−2 4

taking the vector from O to A. In this case $\bar{\mathbf{r}} =$ __$\bar{\mathbf{i}}$ + __$\bar{\mathbf{j}}$ + __$\bar{\mathbf{k}}$.

The determinant defining the cross product ($\bar{\mathbf{r}} \times \bar{\mathbf{F}}$) is now written as:

$$\begin{vmatrix} \bar{\mathbf{i}} & \bar{\mathbf{j}} & \bar{\mathbf{k}} \\ 5 & -2 & 4 \\ 20 & 30 & -60 \end{vmatrix}$$

0 380 190

Expansion of this determinant shows that $\bar{\mathbf{M}}_O =$ __$\bar{\mathbf{i}}$ + __$\bar{\mathbf{j}}$ + __$\bar{\mathbf{k}}$. This result is the same as the one obtained previously.

We next investigate the moment a force exerts about a chosen line or axis. That is, we are concerned with the moment of a force about an established direction. The only unknown to be determined is the magnitude of the moment. Here, we consider the

scalar

moment not as a vector, but as a ____.

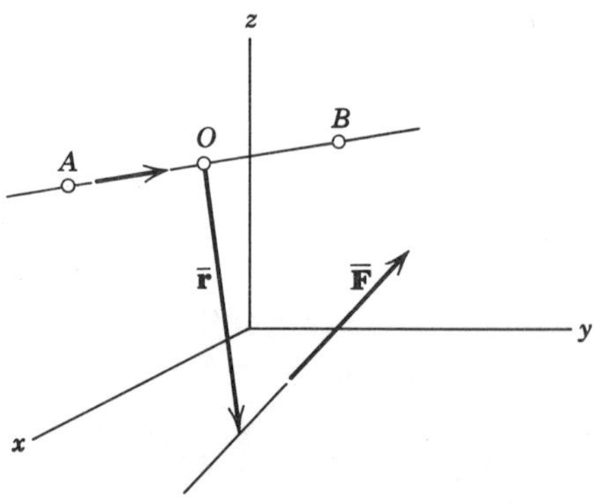

Figure 3.5

In Fig. 3.5, to consider *the moment of the force* $\bar{\mathbf{F}}$ *about the line AB*, we first choose a point O on line AB as the origin of $\bar{\mathbf{r}}$. The moment of $\bar{\mathbf{F}}$ about this point is $\bar{\mathbf{M}}_O = \bar{\mathbf{r}} \times \bar{\mathbf{F}}$. We then obtain the component of this moment in the direction AB by the relation that $M_{AB} = \overline{\mathbf{M}} \cdot \bar{\mathbf{e}}_{AB}$. Thus the moment about AB can be written as $M_{AB} = (\bar{\mathbf{r}} \times \bar{\mathbf{F}}) \cdot$ __. $\bar{\mathbf{e}}_{AB}$

To clarify this definition, we consider the torque exerted by $\bar{\mathbf{F}}$ in Fig. 3.6 (through some connection not

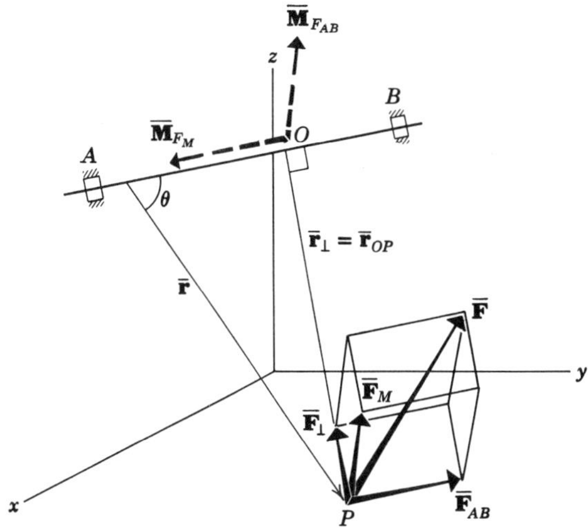

Figure 3.6

shown) on the shaft fixed on line AB. Let the force $\bar{\mathbf{F}}$ be divided into three orthogonal (mutually perpendicular) components. Now consider the moment of each of these components about point O on line AB as a component of the resultant moment of $\bar{\mathbf{F}}$ about point O. This procedure is suggested by the principle of moments referred to earlier. Take the components of the force $\bar{\mathbf{F}}$ in the following manner: take $\bar{\mathbf{F}}_{AB}$ to be the component of $\bar{\mathbf{F}}$ parallel to the line AB. Take the component $\bar{\mathbf{F}}_\perp$, perpendicular to the line AB and intersecting it. Also we define the point O as the intersection of this component with AB. The third and remaining component $\bar{\mathbf{F}}_M$ is perpendicular to AB and to $\bar{\mathbf{F}}_\perp$. Let us examine separately, the moments about O and the line AB of these components. The force $\bar{\mathbf{F}}_\perp$ passes through line AB at the point O. Therefore the moment of $\bar{\mathbf{F}}$ about point O is ___. Thus the compo- zero nent of this moment acting in the direction of the line AB is ___.

zero

Now consider the moment of $\bar{\mathbf{F}}_{AB}$ about point O. By definition, the moment of a force about a point is a vector perpendicular to the plane of the point and the force. Thus the moment $\bar{\mathbf{M}}_{FAB}$ about point O of force $\bar{\mathbf{F}}_{AB}$ is perpendicular to the line AB and does not have a component in the direction AB. We conclude that, about the line AB, the moment of the force $\bar{\mathbf{F}}_{AB}$ equals ___.

zero

That the forces $\bar{\mathbf{F}}_{\perp}$ and $\bar{\mathbf{F}}_{AB}$ produce no moment on the shaft is intuitive since the effect of $\bar{\mathbf{F}}_{\perp}$ is only to move the shaft sideways and the effect of $\bar{\mathbf{F}}_{AB}$ is to slide the shaft through its bearings. Neither force tends to rotate the shaft.

Therefore, which of the three components of $\bar{\mathbf{F}}$ produces a moment about the line AB? ___.

$\bar{\mathbf{F}}_M$ is the only component

Recall that we chose the component $\bar{\mathbf{F}}_M$ perpendicular to the line OP. Since AB is perpendicular to both $\bar{\mathbf{F}}_M$ and OP, the moment of $\bar{\mathbf{F}}_M$ about point O (given by $\bar{\mathbf{r}}_{OP} \times \bar{\mathbf{F}}_M$) would have to be ___ to the line AB.

parallel

Since the moment of $\bar{\mathbf{F}}_M$ about O is parallel to line AB, its component in the AB direction is its magnitude. Therefore the moment of $\bar{\mathbf{F}}_M$ about line AB is the magnitude of $\bar{\mathbf{F}}_M$ times the perpendicular distance (OP) from the line of action of $\bar{\mathbf{F}}_M$ to the line AB or $M_{AB} = F_M r_{\perp}$. This result can be shown to agree precisely with our formal definition. The right side of the definition $M_{AB} = (\bar{\mathbf{r}}_{OP} \times \bar{\mathbf{F}}) \cdot \bar{\mathbf{e}}_{AB}$ is called a scalar triple product. It can be shown that the order of operations of the scalar triple product may be rearranged without changing its value (so long as cyclic order is maintained). We then rewrite this expression as $M_{AB} = \bar{\mathbf{F}} \cdot (\bar{\mathbf{e}}_{AB} \times \bar{\mathbf{r}}_{OP})$. The cross product $\bar{\mathbf{e}}_{AB} \times \bar{\mathbf{r}}_{OP}$ produces a vector perpendicular to AB and to $\bar{\mathbf{r}}$ and thus parallel to the force component ___.

$\bar{\mathbf{F}}_M$

From Fig. 3.6 we see that the magnitude of this vector equals ___. And this is also the magnitude of the vector ___.

$r \sin \theta$

$\bar{\mathbf{r}}_{\perp}$

The scalar product of $\bar{\mathbf{F}} \cdot \bar{\mathbf{e}}_{AB} \times \bar{\mathbf{r}}$ can be thought of as the component of $\bar{\mathbf{F}}$ in the $\bar{\mathbf{e}}_{AB} \times \bar{\mathbf{r}}$ direction multiplied by the magnitude of $\bar{\mathbf{e}}_{AB} \times \bar{\mathbf{r}}$. Write the scalar product in this fashion and show that $F_M r \sin \theta = F_M r_{\perp}$.

One final point to be mentioned concerns the algebraic sign of the moment of a force about a line. With the unit vector $\bar{\mathbf{e}}_{AB}$ we have assigned a positive direction to line *AB* which, according to the right-hand rule, corresponds to a particular rotation direction about the axis *AB*. Thus, when the moment of a force about this line has a positive sign, we know that the shaft has a tendency to rotate in the appropriate direction given by the unit vector $\bar{\mathbf{e}}_{AB}$. If our moment has a negative sign, we conclude that the rotation would be in the ____________ direction about the

same/opposite

opposite

axis. Thus, if the moment is negative and you sight along the shaft from *A* to *B*, this moment tends to rotate the shaft ____________________.

clockwise/counterclockwise

counterclockwise

It is always wise to look carefully at any given problem to determine the easiest way of solving it. For example, we have seen that if a force is readily divided into components parallel, intersecting, and perpendicular to an axis then an intuitive, scalar approach will find the moment of this force about the axis quite simply. However, if the force is given in general three-dimensional terms, the cross-product approach using vectors is probably the simplest. If a given force is either parallel or perpendicular to the axis of interest, then you ______________ use the

should/should not

should not

general cross product approach.

We have just seen that the moment of a force about a line is the component of the moment of that force about some point on the line in the direction of the line. Conversely, if we find the moment of a given force about each of three orthogonal axes at a point, we have found the three components of the ______ of the force about that point.

moment

In some cases the simplest way to find the moment of a force about a given point is to apply this procedure, i.e., find the moment of the force about three orthogonal axes through that point and then ______________ these moments or components.

add, sum (vectorially)

An example that can be treated in a number of ways is shown in Fig. 3.7. The shaft *AB* is parallel to the *x* axis and intersects the *z* axis at the point $z = 3$.

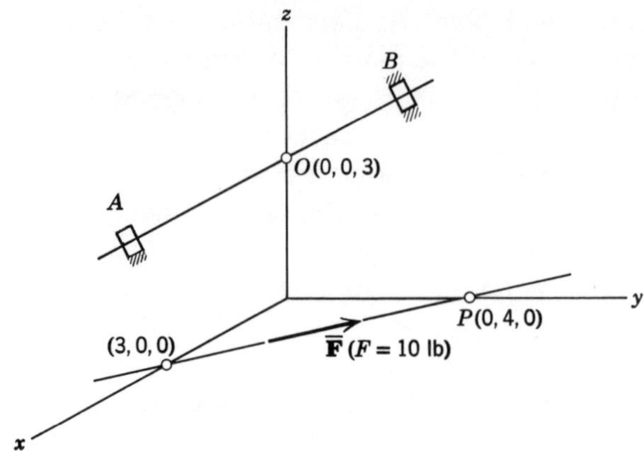

Figure 3.7

We seek the moment the force $\overline{\mathbf{F}}$ exerts on the shaft *AB*. First, we choose a point on *AB* and a point on the line of action of the force $\overline{\mathbf{F}}$. Arbitrarily we choose (0,0,3) as point *O* on *AB* and (0,4,0) as point *P* on $\overline{\mathbf{F}}$. The vector $\overline{\mathbf{r}}_{OP}$ is __$\overline{\mathbf{i}}$ + __$\overline{\mathbf{j}}$ + __$\overline{\mathbf{k}}$.

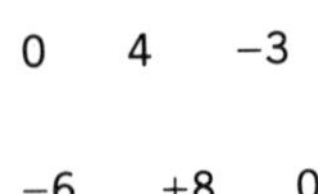

0 4 −3

Write the component expression for the force $\overline{\mathbf{F}}$ as $\overline{\mathbf{F}}$ = ___$\overline{\mathbf{i}}$ + ___$\overline{\mathbf{j}}$ + __$\overline{\mathbf{k}}$ pounds.

−6 +8 0

With the vector cross product we can write the determinant for $\overline{\mathbf{M}}_O$ as $\overline{\mathbf{M}}_O = \begin{vmatrix} \overline{\mathbf{i}} & \overline{\mathbf{j}} & \overline{\mathbf{k}} \\ 0 & 4 & -3 \\ -6 & 8 & 0 \end{vmatrix}$. Expansion of this determinant yields $\overline{\mathbf{M}}_O$ = ____$\overline{\mathbf{i}}$ + ____$\overline{\mathbf{j}}$ + ____$\overline{\mathbf{k}}$.

24 18
24

To determine the component of this moment in the direction of *AB*, we must define a unit vector parallel to the direction of *AB*. Considering the direction from *A* to *B*, as positive, the vector is $\overline{\mathbf{e}}_{AB}$ = _____$\overline{\mathbf{i}}$.
−/+

minus (−)

Therefore, $\overline{\mathbf{M}}_O \cdot \overline{\mathbf{e}}_{AB}$ = ___________.

−24 foot pounds

The minus sign shows that the force $\overline{\mathbf{F}}$ tends to rotate the shaft in the direction shown in Fig. 3.8 ___. Fig. 3.8*b* corresponds to the ___________
a/*b* negative/positive
$\overline{\mathbf{e}}_{AB}$ direction.

a positive

An alternate way of computing this moment is to divide the force $\overline{\mathbf{F}}$ into components $\overline{\mathbf{F}}_x$ and $\overline{\mathbf{F}}_y$ as in Fig. 3.9. Since the component $\overline{\mathbf{F}}_x$ is _____ to the line of *AB* it has _____ moment about the axis *AB*.

parallel
zero, no

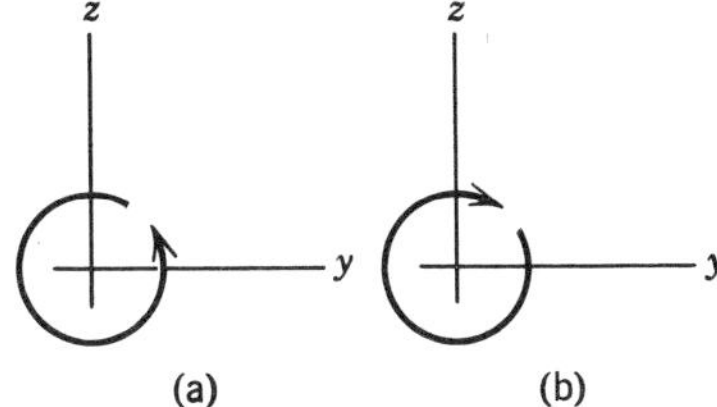

Figure 3.8

The line of action of $\overline{\mathbf{F}}_y$ is perpendicular to *AB*, and the distance between these lines is three feet.

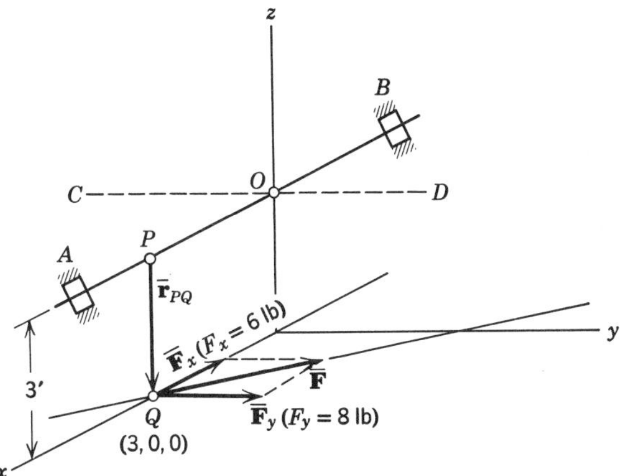

Figure 3.9

The magnitude of the moment of $\overline{\mathbf{F}}_y$ is thus the magnitude 8 pounds multiplied by the 3-foot distance, or 24 foot-pounds. The direction of the vector $\overline{\mathbf{r}}_{PQ} \times \overline{\mathbf{F}}_y$ is parallel to the line ___ and positive from __ to __. AB B A

The answer thus obtained is the same as that obtained using direct vector multiplication.

Let us also use the example of Fig. 3.9, to demonstrate that the components of the moment of the force about point *O* can be obtained by finding the moments of the force about three perpendicular axes through *O*. We have determined that the moment about the axis *AB* has the direction of Fig. 3.8*a* and a magnitude of 24 foot-pounds. Thus the *x* component of $\overline{\mathbf{M}}_O$ is written as ___ foot-pounds. $+24\overline{\mathbf{i}}$

To find the y component we consider the moment about the line CD (Fig. 3.9). Again we divide $\bar{\mathbf{F}}$ into
parallel components $\bar{\mathbf{F}}_x$ and $\bar{\mathbf{F}}_y$. Since $\bar{\mathbf{F}}_y$ is ______ to line CD, it
zero has ___ moment about CD.

The moment of $\bar{\mathbf{F}}$ about CD is only due to the com-
6 ponent $\bar{\mathbf{F}}_x$. The moment has a magnitude of F_x (__
3 pounds) multiplied by the shortest distance (__ feet) between CD and the line of F_x.

Thus the magnitude of the y component of $\bar{\mathbf{M}}_O$
18 is __ foot-pounds. In vector form, this component is
$18\bar{\mathbf{j}}$ written as ____ foot-pounds.

Similarly, the z component of $\bar{\mathbf{M}}_O$ is found by taking the moment of $\bar{\mathbf{F}}$ about the z axis. Again, consider-
$\bar{\mathbf{F}}_x$ ing $\bar{\mathbf{F}}$ as acting at point Q, the component __ does not contribute, since it intersects the z axis.

The moment of $\bar{\mathbf{F}}_y$ about the z axis has a magnitude equal to 8 pounds times 3 feet and can be
$+24\bar{\mathbf{k}}$ written in vector form as ___ foot-pounds.

We see that the moment $\bar{\mathbf{M}}_O$ obtained by this tech-
the same as nique is ________________ the moment ob-
the same as/different from
tained previously by vector multiplication. At times vector multiplication is the simpler way of finding moments and at other times the use of components will be simpler.

We now take up a unique force system that behaves as a pure moment or torque. Such a force system is called a couple and consists of a pair (couple) of forces equal in magnitude with opposite directions and parallel lines of action. Such a pair of forces is shown in Fig. 3.10.

An important property of a couple is that *it has the same moment about every point in space* and thus the moment of a couple about any particular point *does not* depend upon the relative locations of the couple and the point. This means that a couple can be moved about freely without changing its moment with respect to any point in space. To prove this, consider the arbitrary point O in Fig. 3.10. We write the moment of the force $\bar{\mathbf{F}}$ as $\bar{\mathbf{r}}_1 \times \bar{\mathbf{F}}$ and the moment of the force $-\bar{\mathbf{F}}$ as $\bar{\mathbf{r}}_2 \times -\bar{\mathbf{F}}$. Calling the two forces together a couple, the moment of a couple about
$\bar{\mathbf{F}}$ $-\bar{\mathbf{F}}$ point O is then $\bar{\mathbf{M}}_O = \bar{\mathbf{r}}_1 \times$ __ $+ \bar{\mathbf{r}}_2 \times$ __.

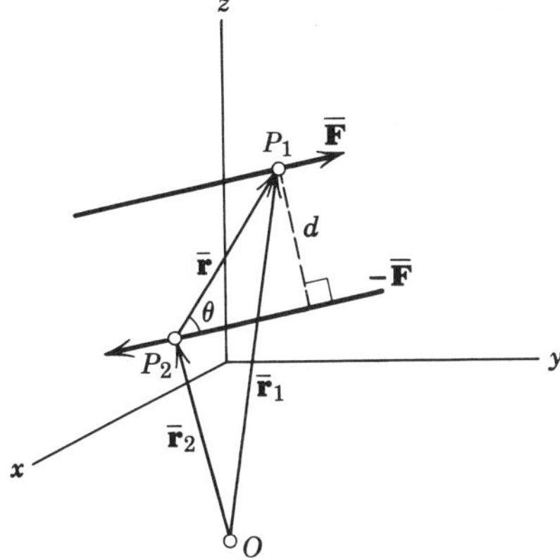

Figure 3.10

This expression can be arranged as $\overline{\mathbf{M}}_O = \overline{\mathbf{r}}_1 \times \overline{\mathbf{F}} - \overline{\mathbf{r}}_2 \times \overline{\mathbf{F}}$. Applying the associative rule for the cross product, we have $\overline{\mathbf{M}}_O = (\overline{\mathbf{r}}_1 - \overline{\mathbf{r}}_2) \times \overline{\mathbf{F}}$. In Fig. 3.10 the vector $(\overline{\mathbf{r}}_1 - \overline{\mathbf{r}}_2)$ is the vector shown and labeled as __. $\overline{\mathbf{r}}$

Thus, when we consider the moment about *any* point O of two forces that are a couple, we need only consider the vectors $\overline{\mathbf{r}}$ and $\overline{\mathbf{F}}$. Since the points P_1 and P_2 that determine $\overline{\mathbf{r}}$ are arbitrary, the moment of a couple about every point in space is the same and is equal to the cross product of any vector $\overline{\mathbf{r}}$, from any point on the line of action of one force to any point on the line of action of the second force, with the second force. From the definition of the cross product, we remember that the term $r \sin \theta$ is simply the ________ ________ distance from the point P_2 to the line of action of $\overline{\mathbf{F}}$. shortest, perpendicular

Since the vectors $\overline{\mathbf{F}}$, $-\overline{\mathbf{F}}$, and $\overline{\mathbf{r}}$ all lie in the same plane, we can also conclude that the vector representing the moment of this couple will be __________ to this plane and to these three vectors. perpendicular

If d is the shortest distance between the lines of action of $\overline{\mathbf{F}}$ and $-\overline{\mathbf{F}}$, then *the magnitude* of the moment vector representing this couple is simply the product of F and __. *d*

We have shown that the moment of a couple about any point is independent of the relative location of the couple and that point. We call the couple vector that represents the moment of the couple a free vector, since its location in space is irrelevant. The couple is completely characterized by the moment vector representing it. We define the equivalence of two couples to mean the same vector represents each couple. In Fig. 3.11 the four couples shown consist of

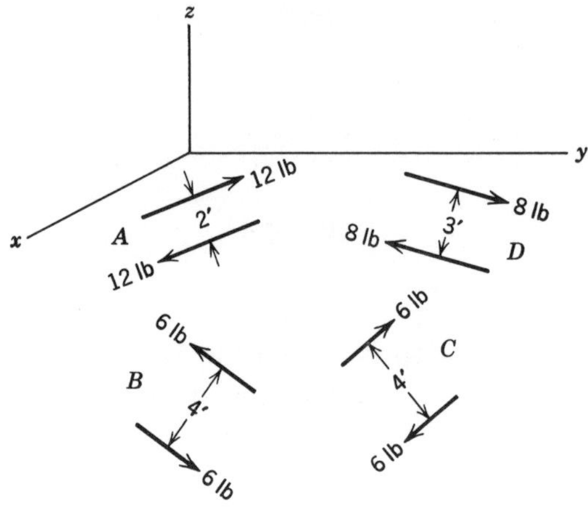

Figure 3.11

forces in the *x-y* plane. Couple *A* is represented by a moment vector of magnitude 24 and written as $-24\bar{\mathbf{k}}$ foot-pounds. From the previous statement we see that
C D couples __ and __ will be equivalent to couple *A*,
B whereas couple __ is not.

It is obvious that the magnitudes of the forces involved in a couple may be changed so long as the product of the magnitude and the distance between the forces remains constant. Yet, for equivalence, not only must the magnitude of the couple vectors be the same but the direction must also be the same.
B For this reason couple __ in Fig. 3.11 is different from the other three.

The directions of the forces comprising a couple are unimportant as long as the moment vector representing the couple remains fixed. Several other characteristics of couples should be mentioned. One of these is that since couples are fully represented by their vectors, the resultant of two couples can be
vectors represented by the resultant of the two ______ that represent these couples.

Since couples are represented by moment vectors, it is a simple matter to determine the moment of a couple about any particular line in space. By analogy with the definition of the moment of a single force about a line, we can conclude that the moment of a couple about a given line will simply be the component of the moment vector of the couple in the direction of the ___. line

As a final comment, note that since couple moments are pure moments, they are often represented simply by a moment vector with no forces included in the diagram.

Having discussed the twisting or torque-producing characteristics of forces that are called moments, we now turn to more general considerations of force systems.

Summary

In general, forces tend to produce rotations and translations of bodies on which they act. The moment of a force is a measure of this tendency to produce rotation. The moment, $\overline{\mathbf{M}}$, of a force $\overline{\mathbf{F}}$ about a point O is defined as the vector product $\overline{\mathbf{r}} \times \overline{\mathbf{F}}$ where $\overline{\mathbf{r}}$ is a vector from O to any point on the line of action of $\overline{\mathbf{F}}$. These vector products may be evaluated from the definition $\overline{\mathbf{M}} = \overline{\mathbf{r}} \times \overline{\mathbf{F}} = rF \sin \alpha \overline{\mathbf{n}}$, where α is the angle between $\overline{\mathbf{r}}$ and $\overline{\mathbf{F}}$ and $\overline{\mathbf{n}}$ is a unit vector normal to the plane defined by $\overline{\mathbf{r}}$ and $\overline{\mathbf{F}}$ and positive in the right-hand sense as $\overline{\mathbf{r}}$ is rotated into $\overline{\mathbf{F}}$. This calculation is usually facilitated by expressing $\overline{\mathbf{r}}$ and $\overline{\mathbf{F}}$ in rectangular components, in which case $\overline{\mathbf{r}} \times \overline{\mathbf{F}}$ is found by expanding the determinant

$$\overline{\mathbf{M}} = \begin{vmatrix} \overline{\mathbf{i}} & \overline{\mathbf{j}} & \overline{\mathbf{k}} \\ r_x & r_y & r_z \\ F_x & F_y & F_z \end{vmatrix}$$

A geometrical interpretation shows that the moment is a vector with magnitude equal to the product of the magnitude of $\overline{\mathbf{F}}$ and the distance from O to the line of action of $\overline{\mathbf{F}}$. This observation often allows a simplification in the calculation of moments by using $\overline{\mathbf{F}}$ in component form.

the moment of the force about any point on the line in the direction of the line. Thus, if the line passes through point O, and is parallel
of the line. Thus, if the line passes through point 0, and is parallel
to the unit vector $\overline{\mathbf{e}}$, the moment of $\overline{\mathbf{F}}$ about the line is $\overline{\mathbf{M}} \cdot \overline{\mathbf{e}} = (\overline{\mathbf{r}} \times \overline{\mathbf{F}}) \cdot \overline{\mathbf{e}}$, where $\overline{\mathbf{r}}$ is as before. The moment of a force about a line is thus a *scalar*, the twisting effect of the force about the line, and has

been shown to equal the product of the shortest distance between $\bar{\mathbf{F}}$ and the line with the component of $\bar{\mathbf{F}}$ perpendicular to the line and to this shortest distance. Clearly if $\bar{\mathbf{F}}$ either parallels or intersects a line, it will have no moment about the line.

A pair of parallel, equal, but oppositely directed, forces are called a couple and have the same moment about every point in space. A couple is a pure moment and is fully characterized by its moment vector. Thus, any two couples with force pairs in parallel planes, the same moment sense, and the same product of force magnitude and separation distance are equivalent.

Problems

(1) If $\bar{\mathbf{A}} = 2\bar{\mathbf{i}} - \bar{\mathbf{j}} - 3\bar{\mathbf{k}}$ and $\bar{\mathbf{B}} = -\bar{\mathbf{i}} + \bar{\mathbf{j}} + 3\bar{\mathbf{k}}$, find: (a) $\bar{\mathbf{A}} \times \bar{\mathbf{B}}$, (b) $\bar{\mathbf{B}} \times \bar{\mathbf{A}}$, and (c) $2\bar{\mathbf{A}} \times 3\bar{\mathbf{B}}$.

(2) A force $\bar{\mathbf{F}} = 6\bar{\mathbf{i}} - 4\bar{\mathbf{j}}$ lb acts at the point $x = 5'$, $y = 1'$, $z = 0$. Determine the moment of the force about the origin of the x-y plane.

(3) A force $\bar{\mathbf{F}} = -3\bar{\mathbf{i}} + 5\bar{\mathbf{j}}$ lb acts at the point $x = 2'$, $y = 1'$, $z = 0$. Determine the moment of the force about point $x = 1'$, $y = 4'$, $z = 1'$.

(4) Determine a unit vector perpendicular to the plane containing $\bar{\mathbf{P}} = \bar{\mathbf{i}} + 2\bar{\mathbf{j}} - 4\bar{\mathbf{k}}$ and $\bar{\mathbf{Q}} = 2\bar{\mathbf{i}} + \bar{\mathbf{j}} - \bar{\mathbf{k}}$.

(5) A body is acted on by a force $\mathbf{F} = \bar{\mathbf{i}} + 2\bar{\mathbf{j}}$ lb applied at the point (0,0,3′). Find the moment of the force about a line through the origin that has its direction given by $\bar{\mathbf{p}} = \bar{\mathbf{i}} + \bar{\mathbf{j}} + \bar{\mathbf{k}}$.

(6) In addition to the couple $\bar{\mathbf{C}} = 5\bar{\mathbf{i}} + 12\bar{\mathbf{j}}$ ft-lb, there is a force $\bar{\mathbf{F}} = 6\bar{\mathbf{i}} - \bar{\mathbf{j}}$ applied at the point (2′,0,−1′) acting on some body B. Find the resultant moment about the line in the direction $\bar{\mathbf{p}} = 3\bar{\mathbf{i}} + 4\bar{\mathbf{j}}$ that passes through the origin.

Answers to Problems

Chapter 3

(1) $\overline{\mathbf{A} \times \mathbf{B}} = -3\bar{\mathbf{j}} + \bar{\mathbf{k}}$

$\overline{\mathbf{B} \times \mathbf{A}} = 3\bar{\mathbf{j}} - \bar{\mathbf{k}}$

$\overline{2\mathbf{A} \times 3\mathbf{B}} = -18\bar{\mathbf{j}} + 6\bar{\mathbf{k}}$

(2) $\bar{\mathbf{M}}_o = -26\bar{\mathbf{k}}$

(3) $\bar{\mathbf{M}}_P = 5\bar{\mathbf{i}} + 3\bar{\mathbf{j}} + 4\bar{\mathbf{k}}$

(4) $\bar{\boldsymbol{\lambda}} = \pm \frac{1}{62} (2\bar{\mathbf{i}} - 7\bar{\mathbf{j}} - 3\bar{\mathbf{k}})$

(5) $\bar{\mathrm{M}}_{OL} = -\sqrt{3}$ ft-lb

(6) $\bar{\mathbf{M}}_{OL} = -2.4$ ft-lb

chapter 4

Force Systems

Objectives

Here we examine the concept of the resultant of a force system and of equivalence of force systems in terms of external effect. Simplified resultants of certain systems common to engineering are discussed and an approach to simple distributed force systems is presented. This chapter attempts to provide the insight and tools necessary so that the student can:

1. Explain external effects of force systems and define equivalence of two systems.
2. Define the resultant of a general force system and calculate typical resultants.
3. Identify the special cases of concurrent coplanar and parallel force systems, state the form of the resultant of each, and calculate their resultants in typical engineering problems.
4. Explain the common approach to finding resultants of distributed force systems and determine, by integration, resultants of simple line loads and pressure loads on simple plane areas.

4.1 GENERAL FORCE SYSTEMS

Statics is a study of those characteristics of force systems that produce *external* effects on rigid bodies and particles. By *external* we mean overall translations and rotations but not deformations.

It can be demonstrated that force systems having the same vector sum of forces and the same moment about every point in space produce identical *external* effects on rigid bodies or particles. Thus we *define* systems of forces (and couples) as *equivalent* if they have the same vector sum of forces and the same moment about every point in space. Let us examine this concept of equivalence of force systems in detail.

Consider force $\bar{\mathbf{F}}$ acting through point Q in Fig. 4.1. We seek to replace this force system with an

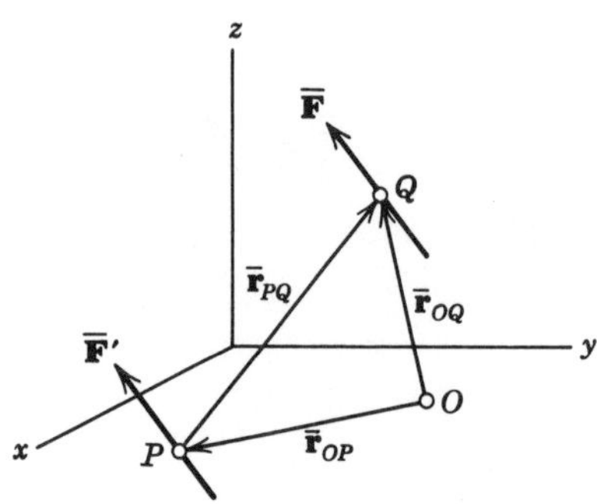

Figure 4.1

equivalent force system consisting of a single force ($\bar{\mathbf{F}}'$) through point P. First, equality of the vector sum of forces requires that the force $\bar{\mathbf{F}}'$ through P have magnitude and direction equal to ___.

$\bar{\mathbf{F}}$

In addition, we require the new system to have equal moments about every point in space. To achieve this we must add a couple. We first write an expression for the moment of the original force system about an arbitrary point, O. The moment of the original force about O is $\bar{\mathbf{r}}_{OQ} \times$ ___.

$\bar{\mathbf{F}}$

However, from Fig. 4.1 it is seen that $\bar{\mathbf{r}}_{OQ}$ can be written as the vector sum of $\bar{\mathbf{r}}_{OP}$ and ___.

$\bar{\mathbf{r}}_{PQ}$

Therefore, the moment of the *original force system* about the point O is equal to $\bar{\mathbf{r}}_{OP} \times \bar{\mathbf{F}} +$ ___ $\times \bar{\mathbf{F}}$.

$\bar{\mathbf{r}}_{PQ}$

Clearly, the moment of force $\bar{\mathbf{F}}'$ about O will be ___ $\times \bar{\mathbf{F}}'$ $(= \bar{\mathbf{r}}_{OP} \times \bar{\mathbf{F}})$.

$\bar{\mathbf{r}}_{OP}$

Since this term is only part of the moment of the original force about O, we must add an additional moment to our new force system so that its moment about point O equals that of the original force system. Suppose we add to $\bar{\mathbf{F}}'$ a couple with a moment equal to $\bar{\mathbf{r}}_{PQ} \times \bar{\mathbf{F}}$. Now the moments of the new and old force systems are ________________. the same

different/the same

Notice, however, that this required couple is simply the moment of the original force $\bar{\mathbf{F}}$ about any point on the line of action of the force $\bar{\mathbf{F}}'$. Thus a single force $\bar{\mathbf{F}}$ through *Q* is *equivalent* to a system consisting of the force $\bar{\mathbf{F}}\,(=\bar{\mathbf{F}}')$ through *P* plus a couple equal to the moment of $\bar{\mathbf{F}}$ about point *P*. If we are concerned with external effects only, a force $\bar{\mathbf{F}}$ through any point *Q* can be replaced by an equivalent force system consisting of a force $\bar{\mathbf{F}}$ through some other point *P* plus an appropriate associated moment. TRUE ____ True
or FALSE ____

The *resultant* of a force system is defined as the *simplest* force system *equivalent* to the original one. By dealing with resultants rather than more general force systems, calculations are often simpler. For example, consider the three-dimensional force system of Fig. 4.2 containing forces and couples. The

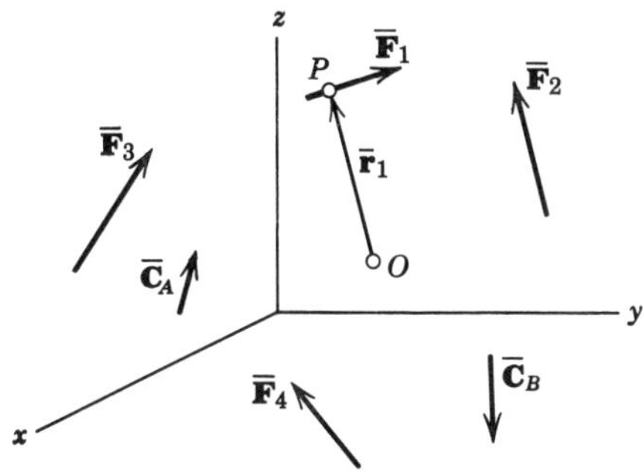

Figure 4.2

resultant sought will involve a single force through the point *O* and an associated couple. We know that $\bar{\mathbf{F}}_1$ can be replaced by an equivalent force system consisting of the force __ through point *O* and an associated couple $\bar{\mathbf{C}}_1$ equal to ____ $\times\ \bar{\mathbf{F}}_1$. $\bar{\mathbf{F}}_1$ $\quad \bar{\mathbf{r}}_{OP} = \bar{\mathbf{r}}_1$

We can also replace each of the forces $\bar{\mathbf{F}}_2$, $\bar{\mathbf{F}}_3$, and $\bar{\mathbf{F}}_4$ by equivalent force systems consisting of that same force through point *O* and its associated ____. couple

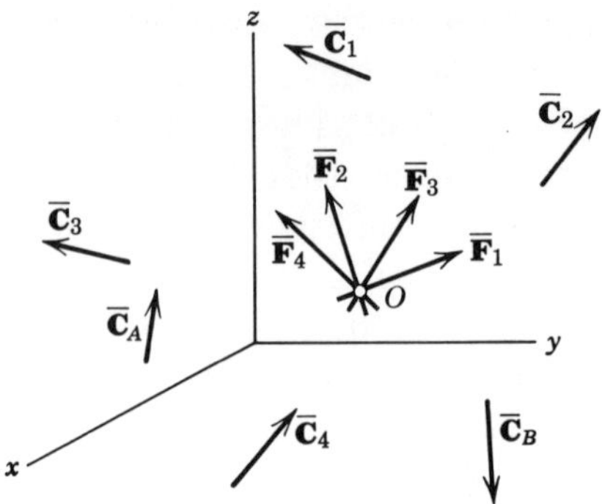

Figure 4.3

In Fig. 4.3 we have replaced the original system
O — by a system of forces concurrent at point ___ and an associated set of couples $\bar{\mathbf{C}}_1$, $\bar{\mathbf{C}}_2$, $\bar{\mathbf{C}}_3$, $\bar{\mathbf{C}}_4$ plus the additional couples present, $\bar{\mathbf{C}}_A$ and $\bar{\mathbf{C}}_B$.

Since all the forces are now concurrent at *O* they
vector — may be summed using ___ addition to find the re-
(type)
force — sultant ___ through point *O*.

Since the couples $\bar{\mathbf{C}}_1$, $\bar{\mathbf{C}}_2$, $\bar{\mathbf{C}}_3$, $\bar{\mathbf{C}}_4$, and $\bar{\mathbf{C}}_A$ and $\bar{\mathbf{C}}_B$ are free vectors, independent of their location, they also can be added. Since the resultant couple is a free
will not — vector, it ___ have a particular line of action.
will/will not

Since point *O* is arbitrary, we conclude that *any* general force system can be replaced by an equivalent system, called its *resultant*. The resultant con-
any — sists of a single force through ___ point
any/a particular
couple — in space and a ___ associated with that point.

The resultant force is unique and is simply the vector sum of the forces involved. However, the resultant moment depends upon the line of action chosen for the resultant force and thus is likely to
vary — ___ for different locations.
vary/be constant

4.2 SOME SPECIAL FORCE SYSTEMS

There are force systems with even simpler resultants. Consider a system of forces that are concurrent at some point in space. If all the forces are moved along their lines of action to the point of concurrency, they may be summed. In this case the resultant of the original system is simply the sum of the forces involved and there is no associated ______________. couple or moment

Hence, a concurrent force system has as its resultant a single force equal to the vector sum of the original forces and acting through the point of __________. concurrency

Consider now a *coplanar* force system. If this force system is also concurrent, then its resultant is a single ______ acting through the point of concurrency force
and lying in the same ______ as the forces of the initial plane
system.

A more general coplanar force system is shown in Fig. 4.4. All the forces lie in the *x-y* plane. First, let

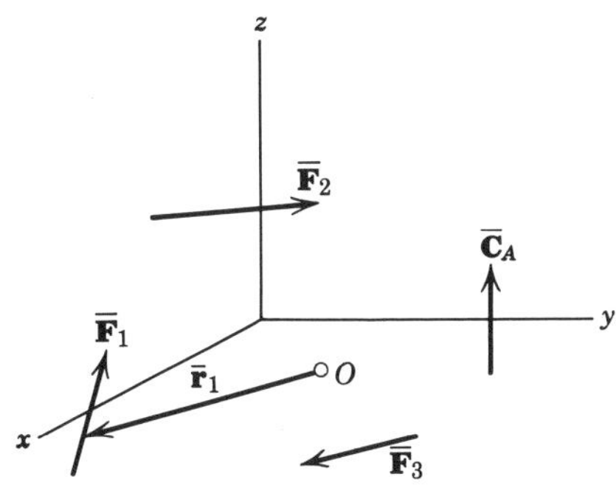

Figure 4.4

us replace this force system by a single force through point *O* in the *x-y* plane and an associated couple. The resultant force through *O* will be the vector ______ of the sum
forces shown and must lie in the ___ plane. *x-y*

The moments about *O* of each of these forces will have the form $\bar{r} \times \bar{F}$ and can be represented by vectors directed ______________________ to the *x-y* perpendicular
parallel/perpendicular
plane.

Similarly, any couple, such as $\bar{C}_A$, formed by two forces lying in the *x-y* plane, can be represented by a vector perpendicular to the *x-y* plane. Thus we can replace the general coplanar force system by an equivalent system consisting of a single force $\bar{\mathbf{F}}_{RO}$ (Fig. 4.5) through a given point, O, and a couple $\bar{\mathbf{C}}_{RO}$ normal to the ____ of the original forces.

plane

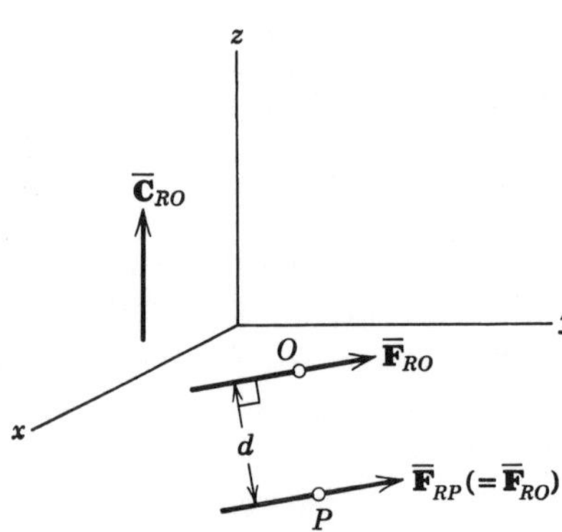

Figure 4.5

Further simplification is possible. In Fig. 4.5 we see the resultant force $\bar{\mathbf{F}}_{RO}$ lying in the x-y plane and the associated moment $\bar{\mathbf{C}}_{RO}$ normal to the x-y plane. We can replace *this* force system with a *single force.* The *single force* $\bar{\mathbf{F}}_{RP}$, equivalent to $\bar{\mathbf{F}}_{RO}$ and $\bar{\mathbf{C}}_{RO}$ must have a magnitude and direction equal to ____.

$\bar{\mathbf{F}}_{RO}$

Where can we place this force so that it will have the same moment about every point in space as the original force system? Suppose we choose a point P in the x-y plane a distance d from the line of action of $\bar{\mathbf{F}}_{RO}$ and require that $(F_{RO})\,(d) = C_{RO}$ (i.e., so that $\bar{\mathbf{r}}_{OP} \times \bar{\mathbf{F}}_{RO} = \bar{\mathbf{C}}_{RO}$). Then the moment about O of the single force through P is precisely ____.

$\bar{\mathbf{C}}_{RO}$

Thus, $\bar{\mathbf{F}}_{RP}$ acting through P ______ equivalent to
is/is not
to $\bar{\mathbf{F}}_{RO}$ and $\bar{\mathbf{C}}_{RO}$.

is

We conclude that a coplanar force system, with nonzero resultant force, can be replaced by a single force acting along a *unique line of action.*

A different situation exists when the resultant force $\bar{\mathbf{F}}_{RO}$ is zero. Here the resultant of the force system can be at most a _____.

couple

Parallel forces are another special case. Figure 4.6 shows a system parallel to the z axis. Couples

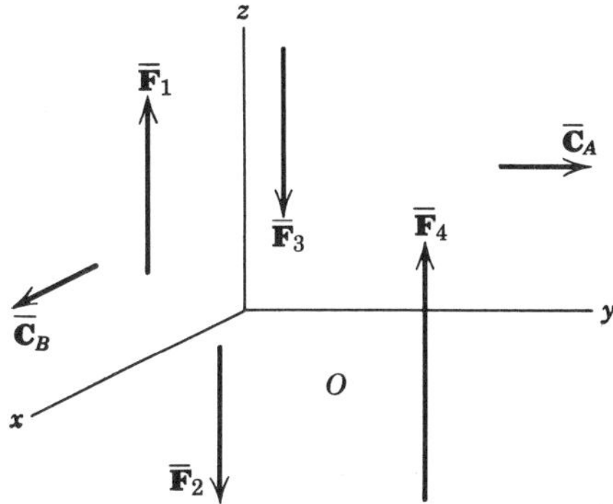

Figure 4.6

consisting of forces parallel to the *z* axis can be represented by vectors lying in the ___ plane. *x-y*

Therefore, this system can be replaced by one consisting of a single force $\bar{F}_R$ parallel to the ___ axis *z* through an arbitrary point *O* and an associated couple. This couple must be parallel to the ___ plane. *x-y*

Effecting this replacement, we obtain the system of Fig. 4.7, where $\bar{F}_{RO}$ acts through *O* and $\bar{C}_{RO}$ is the associated couple. Is there a location for the force $\bar{F}_R$ such that $\bar{F}_R$ alone, without a couple, will be equivalent to the original force system? This line of action of $\bar{F}_R$ can be determined by requiring the moment of $\bar{F}_R$ about *O* to equal $\bar{C}_{RO}$. Refer to Fig. 4.7. If $\bar{r}_{OP} \times \bar{F}_R$

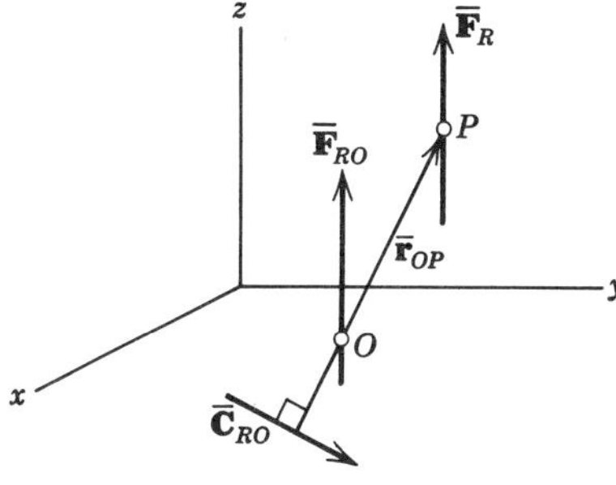

Figure 4.7

$= \bar{C}_{RO}$, then the single force $\bar{F}_R$ through *P* is equivalent to the original force system. This requires $\bar{F}_R$ to pass

normal

through point P on a line through O ________ to
parallel/normal
the moment vector $\bar{\mathbf{C}}_{RO}$.

So, for a *parallel force system* we find the resultant to consist of only a *single force* acting along some unique ___.

line

However, when there is a zero resultant force, i.e., $\bar{\mathbf{F}}_R = 0$, the resultant of a parallel force system consists of no more than a single ____.

couple

Let's outline a general method for finding the resultant of various force systems. First, classify the force system as one of the cases discussed. Then sum the forces involved. If the force sum is zero, we have a resultant equal to a ________.

couple/moment

If the sum of the forces is nonzero, then generally we shall expect a resultant of a force and a couple. For the special cases of concurrent, coplanar, and parallel force systems, the resultant is a single ___ along some ________ line of action.

force

unique/particular

Consider the forces in Fig. 4.8. These forces do

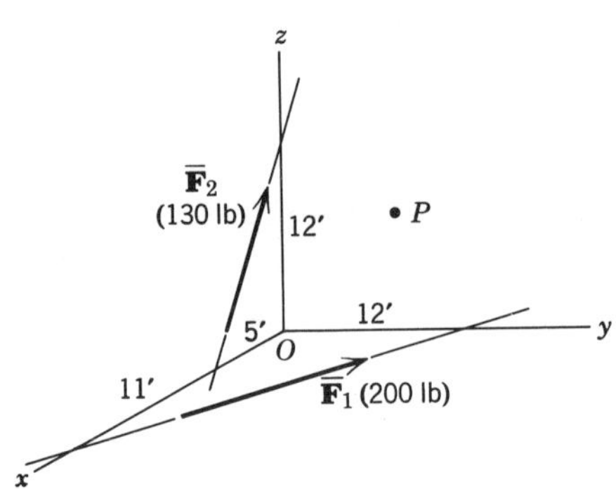

Figure 4.8

not intersect and, therefore, are not ______; they do not lie in the same ____ nor are they parallel.

concurrent

plane

We conclude that this is a general force system and that the resultant will consist of a ___ plus an associated ____.

force

couple

Let us seek a resultant with its force acting through point O. To find $\bar{\mathbf{F}}_R$, sum the forces shown. First, write the forces in component form as $\bar{\mathbf{F}}_1 =$ ___$\bar{\mathbf{i}}$ + ___$\bar{\mathbf{j}}$ + ___$\bar{\mathbf{k}}$ and $\bar{\mathbf{F}}_2 =$ ___$\bar{\mathbf{i}}$ + ___$\bar{\mathbf{j}}$ + ___$\bar{\mathbf{k}}$.

−160,

120, 0 −50, 0, 120

Adding, we find that $\bar{\mathbf{F}}_R$ = ___$\bar{\mathbf{i}}$ + ___$\bar{\mathbf{j}}$ + ___$\bar{\mathbf{k}}$. −210, 120, 120

To find the associated moment, we compute the moments of the forces $\bar{\mathbf{F}}_1$ and $\bar{\mathbf{F}}_2$ about point *O*. Choose vectors connecting the point *O* and the forces as $\bar{\mathbf{r}}_1$ along the *y* axis and $\bar{\mathbf{r}}_2$ along the *z* axis (see Fig. 4.8). We find $\bar{\mathbf{r}}_1$ = ___ and $\bar{\mathbf{r}}_2$ = ___. $12\bar{\mathbf{j}}$ $12\bar{\mathbf{k}}$

Now, compute and add the moments $\bar{\mathbf{M}}_{O1}$ and $\bar{\mathbf{M}}_{O2}$. The couple associated with the resultant force through point *O* is $\bar{\mathbf{C}}_{RO}$ = __$\bar{\mathbf{i}}$ + ___$\bar{\mathbf{j}}$ + ___$\bar{\mathbf{k}}$ ft-lb. 0, −600, 1920

Now we have a force through *O* and an associated couple as the resultant of our force system. To obtain an equivalent force system with the force acting through some other point, say point *P* in Fig. 4.8, we need not repeat the entire process. We simply move the original resultant. The resultant *force* acting through *P* is ______________ the resultant
the same as/different from
force acting through point *O*. The new couple, associated with $\bar{\mathbf{F}}_R$ acting through *P*, equals the couple associated with the force through point *O* *plus* the moment of $\bar{\mathbf{F}}_R$ through point *O* about point __. the same as *P*

Thus the new resultant is $\bar{\mathbf{F}}_R$ through *P* and a couple equal to the sum of $\bar{\mathbf{C}}_{RO}$ and the cross product of $\bar{\mathbf{F}}_R$ with a vector from point __ to point __. *P* *O*

As a second example, we will find the resultant of the coplanar force system of Fig. 4.9.

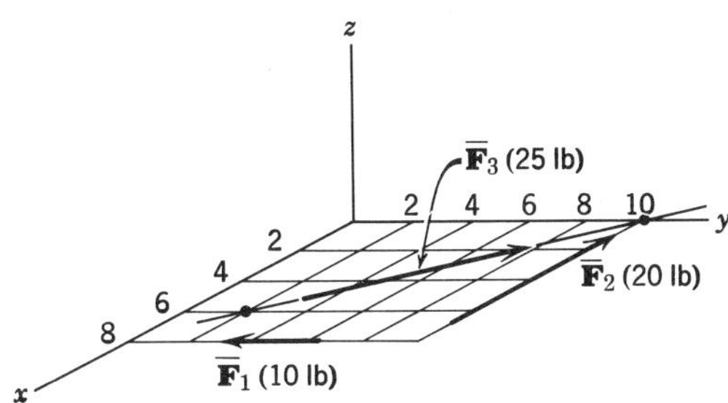

Figure 4.9

Since all of the forces lie in the *x-y* plane, the resultant must lie in the __ plane and all of the moments must be parallel to the __ axis. *x-y* *z*

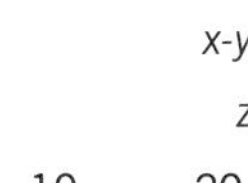

First, write the forces as $\bar{\mathbf{F}}_1$ = __$\bar{\mathbf{i}}$ + ___$\bar{\mathbf{j}}$, $\bar{\mathbf{F}}_2$ = ___$\bar{\mathbf{i}}$ 0, −10 −20,

0 −15, 20 + ___$\bar{\mathbf{j}}$, and $\bar{\mathbf{F}}_3$ = ___$\bar{\mathbf{i}}$ + ___$\bar{\mathbf{j}}$. It is now a simple matter to
−35, 10 find the resultant force $\bar{\mathbf{F}}_R$ = ___$\bar{\mathbf{i}}$ + ___$\bar{\mathbf{j}}$.

We must position this resultant force along some line of action so as to have the same moment about every point in space as the original force system. Taking moments about O, we see that the moment of $\mathbf{F}_1$
+200 is $-80\bar{\mathbf{k}}$ ft-lb and the moment of $\bar{\mathbf{F}}_2$ is ___$\bar{\mathbf{k}}$ ft-lb.

To find the moment of $\bar{\mathbf{F}}_3$, use a simple scalar approach. Consider $\bar{\mathbf{F}}_3$ as acting at point (0, 10, 0) on the y axis. The $\bar{\mathbf{j}}$ component of the force has no moment about point O, since it passes through O. The moment
+150 of $\bar{\mathbf{F}}_3$, produced solely by its $\bar{\mathbf{i}}$ component, is ___ $\bar{\mathbf{k}}$
ft-lb.

The total moment of the original force system about O is the sum of the moments of the individual
270 forces. Thus $\bar{\mathbf{M}}_O$ equals ___ $\bar{\mathbf{k}}$ ft-lb.

With $\bar{\mathbf{M}}_O$ known, we now can find the line of action of $\bar{\mathbf{F}}_R$. Since $\bar{\mathbf{F}}_R$ is not parallel to the x axis, it must intersect the x axis at some point. Let us choose the point of intersection to satisfy the moment requirement. If $\bar{\mathbf{F}}_R$ acts at this point of intersection, the $\bar{\mathbf{i}}$ component of $\bar{\mathbf{F}}_R$ has no moment about O, since it
O passes through point ___.

The point of intersection of $\bar{\mathbf{F}}_R$ with the x axis can by specified by a distance, d, from point O so that the moment of the y component of $\bar{\mathbf{F}}_R$, $\bar{\mathbf{F}}_{Ry}$, will equal $\bar{\mathbf{M}}_O$. The moment of $\bar{\mathbf{F}}_R$ may be written in scalar form as $(d)\,(F_{Ry}) = M_O$. Since M_O equals 270 and F_{Ry} equals 10,
+27 it is clear that d equals ___.

We have found a point on the line of action of the resultant force, $\bar{\mathbf{F}}_R$. Since the resultant of a coplanar
line of action force system is a single force along a unique ___ ___
___ we have found the resultant sought.

parallel The force system shown in Fig. 4.10 is a ___
(type)
force system. Its resultant is either a single ___ acting
couple along a unique line or a ___.

We first find the resultant of the forces $\bar{\mathbf{F}}_2$ and $\bar{\mathbf{F}}_3$. The magnitude of the resultant of these two forces
−30 equals ___$\bar{\mathbf{k}}$ lb.

Since the two forces are coplanar, their resultant must lie in the same plane and intersect the line de-
4 fined by x = ___, z = 0.

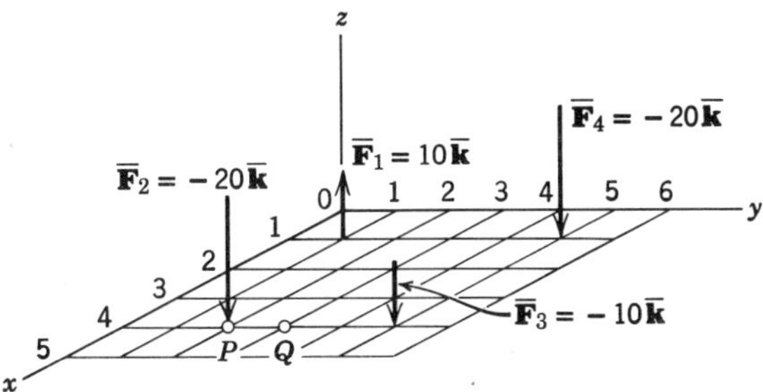

Figure 4.10

The other requirement is that this resultant force have the same moment about every point in space as did the original two forces. Consider the point of intersection of $\bar{F}_2$ with the *x-y* plane, point (4,2,0). The moment of the original force system is solely the moment of force ___ about this point, since $\bar{F}_2$ passes through the point. $\bar{F}_3$

The moment of $\bar{F}_3$ about this point is ____ ft-lb. $-30\bar{i}$

For the resultant force, $-30\bar{k}$ lb, to have the same moment about *P*, it must pass at a distance from *P* of ____. Thus, in Fig. 4.10, the resultant of $\bar{F}_2$ and $\bar{F}_3$ passes through point ___. 1 ft Q

We have now found the resultant of a pair of parallel forces. We can repeat this process, adding one additional force to each resultant obtained and eventually find the resultant of the entire force system. A more direct way, however, is to use the requirement that the moment of the resultant about every point in space must be the same as that of the original force system.

Let's rework the example to demonstrate. The resultant $\bar{F}_R$ of the entire force system, consisting of $\bar{F}_1$, $\bar{F}_2$, $\bar{F}_3$, and $\bar{F}_4$, is $\bar{F}_R =$ ______. $-40\bar{k}$ lb

To locate this resultant force, we first find the moment of the original force system about some reference point. We here choose point *O*. Recall from Chapter 3 that the components of the moment about point *O* are the moments about a set of perpendicular axes through point *O*.

Consider $\bar{F}_1$. It passes a perpendicular distance of 1 ft from the *x* axis and, therefore, has an $\bar{i}$ component of moment about O of (1 ft × 10 lb) $10\bar{\mathbf{i}}$ ft-lb. Similarly, $\bar{\mathbf{F}}_1$ passes 1 ft from the y axis and has a $\bar{\mathbf{j}}$ component of moment about O equal to $-10\bar{\mathbf{j}}$ ft-lb. Thus, $\bar{\mathbf{M}}_{1o}$

$10\bar{\mathbf{i}}$ $-10\bar{\mathbf{j}}$ equals ___ + ___ ft-lb.

Would any of the moments have z components?

No, since the forces are parallel to the z axis. ___. Why? ________________.

Similarly, $\bar{\mathbf{F}}_2$ passes 2 ft from the x axis and has an $\bar{\mathbf{i}}$ component of moment about O of $-40\bar{\mathbf{i}}$ ft-lb. $\bar{\mathbf{F}}_2$

4 passes a distance ___ ft from the y axis and thus has

+80 a $\bar{\mathbf{j}}$ component of moment about point O of ___$\bar{\mathbf{j}}$ ft-lb.

Thus we find that $\bar{\mathbf{M}}_{2o} = 40\bar{\mathbf{i}} + 80\bar{\mathbf{j}}$ ft-lb. Use

$-50\bar{\mathbf{i}} + 40\bar{\mathbf{j}}$ scratch paper to find that $\bar{\mathbf{M}}_{3o}$ equals ________ + ________ ft-lb and that $\bar{\mathbf{M}}_{4o}$ equals ________

$-100\bar{\mathbf{i}} + 20\bar{\mathbf{j}}$ + ________ ft-lb.

Summing, we find that the resultant moment

$-100\bar{\mathbf{i}} + 130\bar{\mathbf{j}}$ $\bar{\mathbf{M}}_{RO}$ of the original force system equals ________ ft-lb.

We must now find the line of action of $\bar{\mathbf{F}}_R$ ($-40\bar{\mathbf{k}}$ lb) that will provide this same moment about the origin. Clearly the line of action must intersect the x-y plane. We identify the point of intersection by a vector $\bar{\mathbf{r}}$ with components $x\bar{\mathbf{i}} + y\bar{\mathbf{j}}$ to be determined. We write $\bar{\mathbf{r}} \times \bar{\mathbf{F}}_R = \bar{\mathbf{M}}_{RO}$, where $\bar{\mathbf{r}} = x\bar{\mathbf{i}} + y\bar{\mathbf{i}}$. This moment is $(x\bar{\mathbf{i}} + y\bar{\mathbf{j}}) \times (-40\bar{\mathbf{k}}) = (-100\bar{\mathbf{i}} + 130\bar{\mathbf{j}})$ where x and y are the coordinates of the point of intersection. This is a vector expression of the fact that $\bar{\mathbf{F}}_R$ must be located so that its moment about the x axis will be $-100\bar{\mathbf{i}}$

$+ 130\bar{\mathbf{j}}$ and about the y axis will be ____.

We can either solve this vector equation directly or take a scalar approach. For insight we choose the latter. For $\bar{\mathbf{F}}_R$ to have a moment about the x axis of magnitude 100, it must be located a distance y from the x axis so that $40y$ equals 100. Choosing the correct sign, we write $-40y = -100$ and find $y = 2.5'$.

$y = 2.5$ This means $\bar{\mathbf{F}}_R$ must act at a distance ____ ft from the x axis.

Similarly, the moment of $\bar{\mathbf{F}}_R$ about the y axis must equal the y component of $\bar{\mathbf{M}}_{RO}$. Calling x the distance of $\bar{\mathbf{F}}_R$ from the y axis, we can write the equation

$40x = 130$ ________.

Solving for x requires $\bar{\mathbf{F}}_R$ to act along the line

3.25 $y = 2.5$ ft; $x =$ ___ ft.

Verify this result directly by expanding and solving the original vector equation.

4.3 DISTRIBUTED FORCE SYSTEMS

Systems with forces distributed continuously over areas or volumes rather than concentrated at certain points are called distributed force systems. To analyze distributed force systems, we first consider the area or volume involved as divided into differential elements. The force acting on each differential element is thus a point force. The sum of these forces over all elements is the resultant force of the system. With this approach, we expect that the resultant of a general three-dimensional distributed force system is a single ____ and an associated ____ and that special cases will yield simpler resultants.

force

couple

Let us apply this method to an example. Consider Fig. 4.11. The flat wall of the tank containing liquid

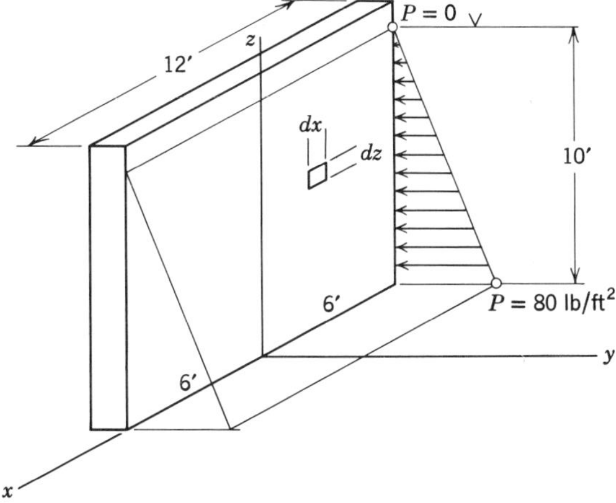

Figure 4.11

is 12 feet wide and the liquid depth is 10 feet. Furthermore, the liquid has a force intensity or pressure at the base of the wall equal to 80 lb/ft². Notice from the units that the product of pressure (force intensity) and area is force. This pressure varies linearly with depth as shown. We seek the resultant of the forces exerted on the wall by the liquid. This resultant intersects the wall at a point called the *center of pres-*

parallel

sure. We can classify the force system as ______ coplanar/ ______, since the forces are in the $-y(-\bar{\mathbf{j}})$ direction. parallel

a single force along some unique line of action

Since the resultant force, $\bar{\mathbf{F}}_R$, is not zero, we expect the resultant of the force system to be ______________________.

To find the magnitude of $\bar{\mathbf{F}}_R$, we sum the forces acting on all the area elements of the wall. Consider the differential force $d\bar{\mathbf{F}}_R$ acting on a differential area element $dxdz$.

The differential force $d\bar{\mathbf{F}}_R$ equals the product of the pressure and the differential area, i.e., $d\bar{\mathbf{F}}_R = P \times dx \times dz\,(-\bar{\mathbf{j}})$. The total force thus is the integral of $d\bar{\mathbf{F}}_R$ over the area, $\bar{\mathbf{F}}_R = -\bar{\mathbf{j}}\int_{\text{area}} Pdxdz$. The pressure, P, varies with the liquid depth and can be written as a function of the __ coordinate.

z

Since the pressure is 80 lb/ft² at $z = 0$ and zero at $z = 10$ ft, we can express the pressure as $P(z) = \dfrac{80(10-z)}{10}$ lb/ft². Thus the differential force on the area element is $-Pdxdz\bar{\mathbf{j}}$, which equals ______________.

$-80\,\dfrac{(10-z)}{10} \times dx \times dz\bar{\mathbf{j}}$ lb

The final integral for $\bar{\mathbf{F}}_R$ is

$$\int_{z=0}^{z=10}\left[-80\,\frac{(10-z)}{10}\,\bar{\mathbf{j}}\int_{x=-6}^{x=6} dx\right] dz$$

Integration yields the resultant force $\bar{\mathbf{F}}_R =$ ____ lb.

$-4800\bar{\mathbf{j}}$

The resultant is located by considering moments. First, consider moments about the z axis. We can conclude by symmetry that there is ______ resultant some/no moment about the z axis due to the liquid and, therefore, that the resultant must intersect the __ axis.

no

z

We need to find the point of intersection of $\bar{\mathbf{F}}_R$ with the z axis. To do this, first find the resultant moment of the original force system about the x axis. We can then locate $\bar{\mathbf{F}}_R$ to have the same moment about the x axis as ______________________.

the original force system

The differential force acting on each differential

area is $d\bar{\mathbf{F}}_R = -PdA\bar{\mathbf{j}} = \left[-80\,\frac{(10-z)}{10}\right] dxdz\bar{\mathbf{j}}$. The moment about the x axis of each of these differential forces is the product of the force magnitude and its distance from the x axis. The distance in this case is z and we can write $dM_x = zdF_R =$ ______________. $\qquad z\left[80\,\frac{(10-z)}{10}\right] dxdz$

The total moment is the integral of dM_x from $x = -6$ to $+6$ and from $z = 0$ to $z =$ __. $\qquad$ 10

On scratch paper, compute this resultant moment. M_x equals ________ ft-lb. $\qquad 16.0 \times 10^3$

The resultant force $\bar{\mathbf{F}}_R$ must act through the point z^* where $z^*F_R = M_x$. Thus, $\bar{\mathbf{F}}_R$ must act through the point $x = 0$, $y = 0$, $z^* =$ _____. $\qquad$ 3.33 ft

We have *the resultant* of this distributed force system, since $\bar{\mathbf{F}}_R$ and its unique line of action are known. Recall that point $x = 0$, $y = 0$, $z = 3.33'$ on the wall through which $\bar{\mathbf{F}}_R$ acts is called the _____ __ ______. $\qquad$ center of pressure

This concludes our general discussion of force systems. In the next chapter, we will consider some useful properties of areas and masses associated with first moments.

Summary

Two force systems which have the same vector sum of forces and the same moment about every point in space will produce the same external effect on a rigid body; thus they are called equivalent. We define the resultant of a force system as the simplest force system equivalent to it. In general, the resultant of a force system is a single force plus an associated couple. In several important special cases it may be simpler, however. For example, a concurrent force system has a single force as a resultant; coplanar and parallel force systems have a single force along a unique line of action *or* a couple as resultants. Distributed forces are treated by first considering the force acting on each differential element and then summing forces and moments by integration. As with sets of concentrated forces, distributed force systems have a resultant consisting of a single force and a couple in the general case and of either a single force or a couple for the special cases listed above.

Problems

(1) Replace the force $\bar{\mathbf{F}}_2$ in Fig. P4.1 by a force at P and a couple. What is the moment of the couple?

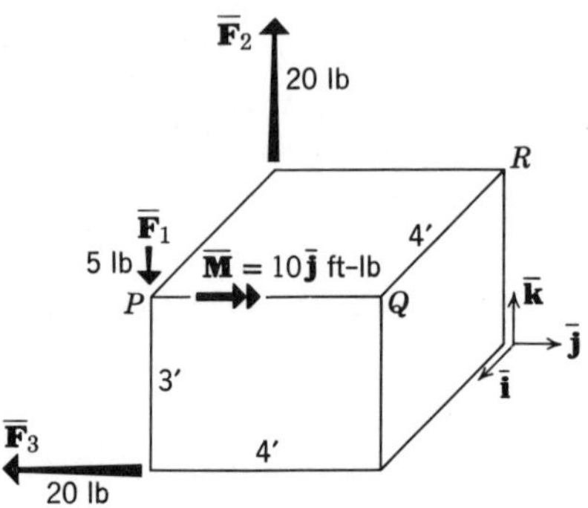

Figure P4.1

(2) In Fig. P4.1, replace the force $\bar{\mathbf{F}}_1$ and the couple, whose moment is $\bar{\mathbf{M}}$, by a single force, say $\bar{\mathbf{F}}_1'$. Where does the line of action of $\bar{\mathbf{F}}_1'$ intersect the PQR plane?

(3) Consider the system of forces in Fig. P4.1. (a) Determine the sum of the forces. (b) What is the sum of the moments of the forces about the point P? What are the units of the moments calculated? (c) Is the system statically equivalent to the system of forces acting at P, consisting of $\bar{\mathbf{F}} = -20\bar{\mathbf{j}} + 15\bar{\mathbf{k}}$ lb and the couples whose moments are $\bar{\mathbf{M}}_1 = 90\bar{\mathbf{i}}$ ft-lb and $\bar{\mathbf{M}}_2 = -60\bar{\mathbf{i}}$ ft-lb?

(4) Replace the concurrent system of forces in Fig. P4.2 with an equivalent single force.

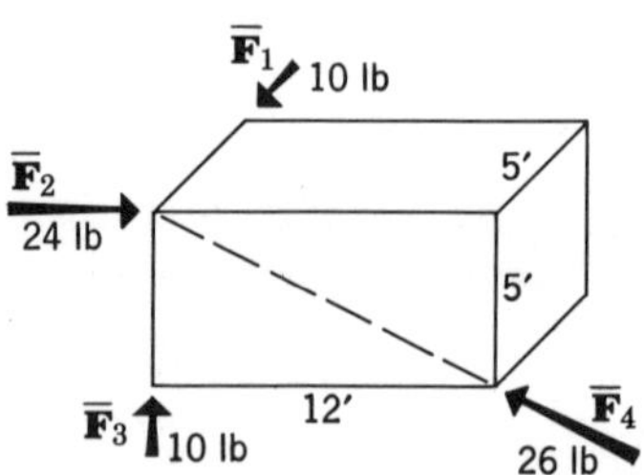

Figure P4.2

(5) Replace the coplanar system of forces in Fig. P4.3 by an equivalent single force.

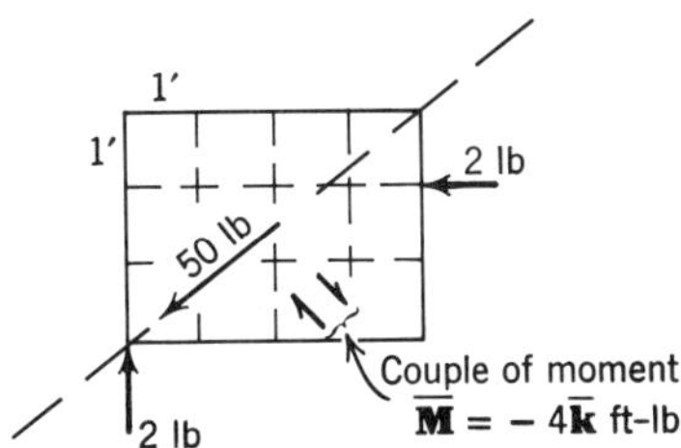

Figure P4.3

(6) Replace the parallel noncoplanar system of forces in Fig. P4.4 by an equivalent single force.

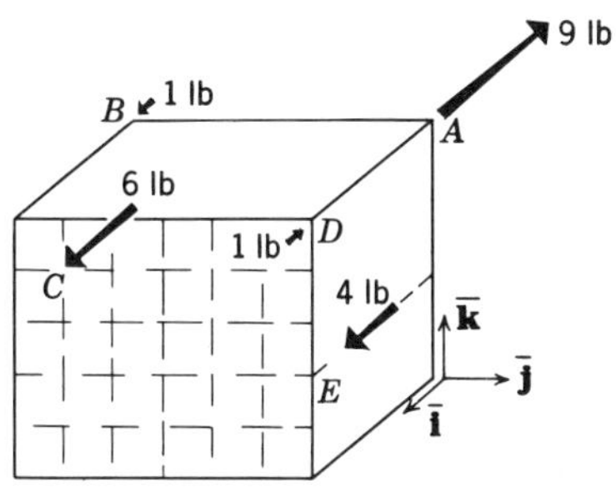

Figure P4.4

(7) In Problems (4), (5), and (6), all equivalent force-couple systems have a common property. What is it?

(8) Replace the distributed load due to the static homogeneous liquid acting on the 10′ long quarter cylinder in Fig. P4.5 with a single statically equivalent force. What are the x and y components of this equivalent force?

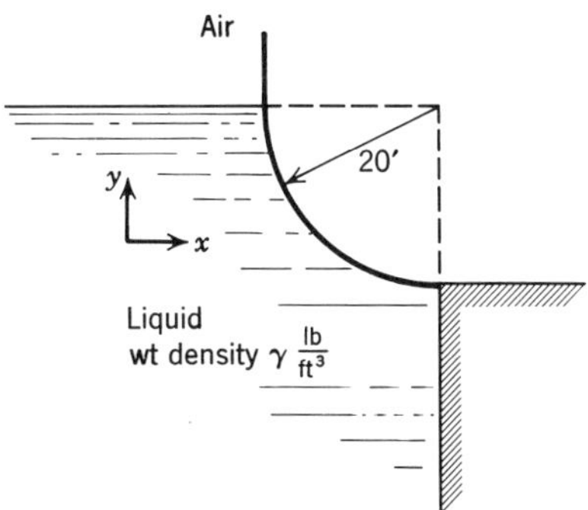

Figure P4.5

Answers to Problems

Chapter 4

(1) $\overline{\mathbf{F}} = 20\overline{\mathbf{k}}$

$\overline{\mathbf{M}}_p = 80\overline{\mathbf{j}}$

(2) $(x,y) = (2,0)$

(3) a. $\Sigma\overline{\mathbf{F}} = -20\overline{\mathbf{j}} + 15\overline{\mathbf{k}}$

b. $\Sigma\overline{\mathbf{M}} = -60\overline{\mathbf{i}} + 90\overline{\mathbf{j}}$

c. Yes

(4) $\overline{\mathbf{R}} = 20\overline{\mathbf{j}} + 10\overline{\mathbf{k}}$

(5) $\overline{\mathbf{R}} = -42\overline{\mathbf{i}} - 28\overline{\mathbf{j}}$

(6) $\overline{\mathbf{R}} = \overline{\mathbf{i}}$, point of intersection on grid face: $(x,y) = (30,13)$

(7) $\overline{\mathbf{R}} \perp \overline{\mathbf{M}}$

(8) $\overline{\mathbf{R}} = 2000\gamma\overline{\mathbf{i}} + 1000\gamma\pi\overline{\mathbf{j}}$

yes

chapter 5

First Moments—Centroids, Centers of Mass, and Gravity

Objectives

Here the concepts of centroid and center of mass and gravity are defined and examined. At the conclusion of this chapter the student should be able to:

1. Define the center of mass and gravity of collections of discrete particles and of continuous bodies and the centroids of areas and volumes.
2. Calculate these properties for simple configurations common to engineering.
3. Explain those conditions of symmetry that restrict the location of centroids and centers of mass and gravity and correctly apply these conditions to simplify typical problems.
4. Explain the composite system approach and apply it to determine various properties of bodies and shapes consisting of collections of elementary shapes.

In the analysis of systems that have material distributed in space, it is often convenient to use certain average or characteristic properties. Here we will investigate some properties associated with moments about axes. We begin with a special kind of distributed force system; that is, the gravity forces acting on the collection of particles shown in Fig. 5.1. Consider the magnitude and direction of the resultant of this force system. Since all the forces are parallel, the resultant force must be in the same direction as the individual forces and must have a magnitude equal to the sum ______________. $W_1 + W_2 + W_3 + W_4$

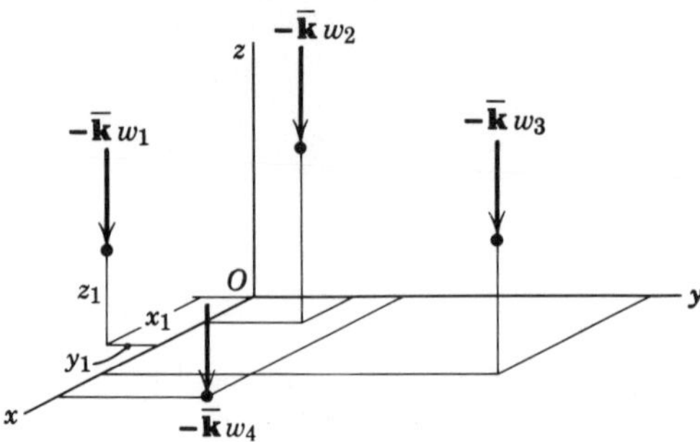

Figure 5.1

Thus, for this system $\bar{\mathbf{F}}_R = (w_1 + w_2 + w_3 + w_4)(-\bar{\mathbf{k}})$. Of course, F_R equals $w_1 + w_2 + w_3 + w_4$ which is W, the total particle weight. Now, where is the line of action of $\bar{\mathbf{F}}_R$?

Since the forces are parallel to the z axis, the location of the resultant is specified by its intersection with the x-y plane. To find the x coordinate, x^*, of the point of intersection, we must find a value of x such that the moment of the resultant about the y axis, $(\bar{\mathbf{i}}x^*) \times \bar{\mathbf{F}}_R$, equals the sum of the moments of the individual particle forces about the y axis. In scalar notation, we write this as $x^*W = \sum_i x_i w_i$. Similarly we find the y coordinate, y^*, of the line of action of the resultant force system using $y^*W = \sum_i$ ____.

$w_i y_i$

Consider the following possibility. Suppose the direction of the gravitational force were changed so that it acts parallel to the y axis rather than the z axis. Such a situation is shown in Fig. 5.2. Here the geome-

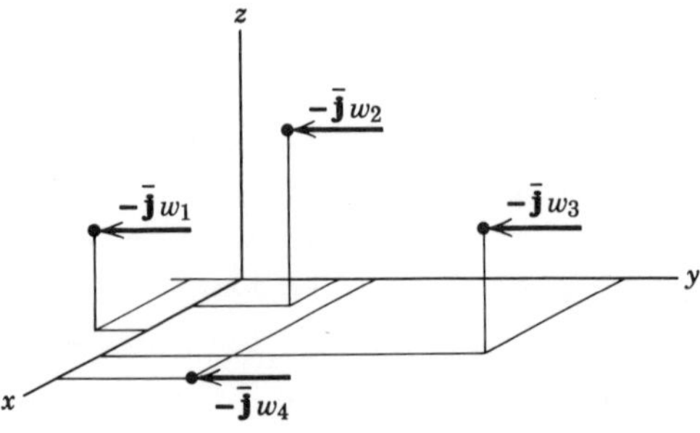

Figure 5.2

try is unchanged from the previous case. The location of the line of action of the resultant force is specified by its intersection with the x-z plane. Since w_1, w_2, w_3, and w_4 are unchanged, the resultant force $\bar{\mathbf{F}}_R$ with magnitude equal the total weight of the system, W, is ________ the previous case.
the same as/different from

the same as

Where is the line of action of $\bar{\mathbf{F}}_R$? The x coordinate of this point, x^*, is found from the expression $x^*W = \sum_i x_i w_i$. Since this expression is ________
identical with/
different from
the one obtained with gravity in the z direction, we conclude that the value of x^* is unchanged.

identical with

To find the z coordinate of the intersection, z^*, we write $z^*W =$ ____.

$\sum_i z_i w_i$

Next consider the gravitational force as acting parallel to the x axis. Again, the point of intersection of the resultant force with the y-z plane, ($y=y^*$, $z=z^*$) is determined from the expressions $y^*W = \sum_i y_i w_i$ and $z^*W = \sum_i z_i w_i$. Thus we must obtain ________
the same/different
values for y^* and z^* as in the previous two cases.

the same

Indeed, if the geometry is fixed and the weights of the individual particles are constant, there exists *one* point (coordinates __, __, __) through which the resultant gravitational force will act regardless of the orientation of the particles relative to gravity. This important point is called the *center of gravity* of the system.

x^* y^* z^*

We can find the center of gravity for a set of discrete particles in a straightforward, though tedious, manner. A more common situation involves continuous distributions of mass. *To treat solid bodies, we consider them as consisting of a number of differential elements of mass, all of which are treated as particles.* Consider the continuous body in Fig. 5.3. The element of mass with sides dx, dy, and dz is located by the position vector $\bar{\mathbf{r}}$. The volume dV of the element equals the product of dx, dy, and dz. If the material has mass density ρ per unit volume, its weight per unit volume is ρg and the weight of the differential element, dW, equals __ dV.

ρg

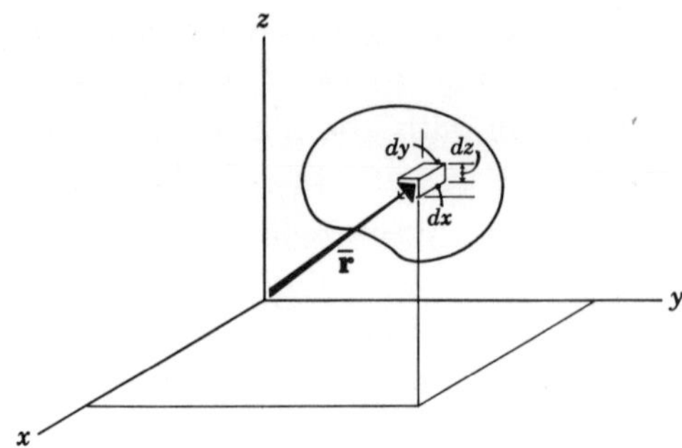

Figure 5.3

The total weight of the body is obtained by integration; for the entire volume $W = \int_{vol} \rho g \, dV$. Therefore, regardless of the direction of the gravitational force, the magnitude of $\overline{\mathbf{F}}_R$ will equal ________.

$W = \int_{vol} \rho g \, dV$

We wish to find that point, called the *center of gravity* of the continuous body, through which $\overline{\mathbf{F}}_R$ passes for any orientation with respect to gravity. For a collection of particles, this is equivalent to finding the point, located by the vector $\bar{\mathbf{r}}^*$ with components x^*, y^*, and z^*, such that Wx^* equals $\sum_i x_i w_i$ and similarly for y^* and z^*. Thus, for a continuously distributed system, $x^*W = \int_{vol} x \, dW = \int_{vol} x\rho g \, dV$. Similarly, write $y^*W =$ ______ and $z^*W =$ ______.

$\int_{vol} y\rho g dV \quad \int_{vol} z\rho g dV$

We shall postpone the evaluation of such integrals. Observe at this point that we have integral expressions for the location of that point in space called the *center of gravity* of a continuous body.

A concept closely related to center of gravity is *center of mass*. To understand this concept, recall that in discussing center of gravity, we considered each element as having mass ρdV and weight $\rho g \, dV$. We thus write the total mass M of a body as $M = \int_{vol} dM = \int_{vol}$ ___.

$\rho \, dV$

We *define* the center of mass of a system as being that point with coordinates x^*, y^*, and z^* for which $x^*M = \int_{vol} x\rho dV$, $y^*M =$ ______, and $z^*M =$ ______.

$\int_{vol} y\rho dV \quad \int_{vol} z\rho \, dV$

Note that if the gravitational acceleration, g, is

constant, the center of gravity ________ coincide with the center of mass of the system.
will/will not

will

The principal task in finding centers of mass and gravity is the computation of certain volume integrals. Since ρ and g are often constant, we frequently will need to compute integrals such as $\int_{\text{vol}} dV$ and $\int_{\text{vol}} x\,dV$. The frequency of this need has led to the concept of the *centroid of a volume.* The centroid, a geometric concept, is defined as that point with coordinates x^*, y^*, and z^*. Here $x^*V = x^* \int_{\text{vol}} dV = \int_{\text{vol}} xdV$. Similarly, the other coordinates are defined by the relations y^* ____ = ____ and z^* ____ = ____.

$\int_{\text{vol}} dV \quad \int_{\text{vol}} ydV$

$\int_{\text{vol}} dV \quad \int_{\text{vol}} zdV$

This definition, and those of center of mass and gravity, can be represented in an equivalent vector form. For example, we define the location of the centroid by a position vector $\bar{\mathbf{r}}^*$ such that $\bar{\mathbf{r}}^* \int_{\text{vol}} dV = \int_{\text{vol}} \bar{\mathbf{r}}dV$ where $\bar{\mathbf{r}}$ is the position vector to each differential element.

Another important geometrical property is the *centroid of an area.* Consider the area shown in Fig. 5.4. We write $dA = dxdy$ and determine the coordinates x^*, y^* of the centroid of this area from the relations $x^*A = \int_{\text{area}} x\,dA$ and $y^*A = \int_{\text{area}} ydA$. Of course, A, the total area, equals ________.

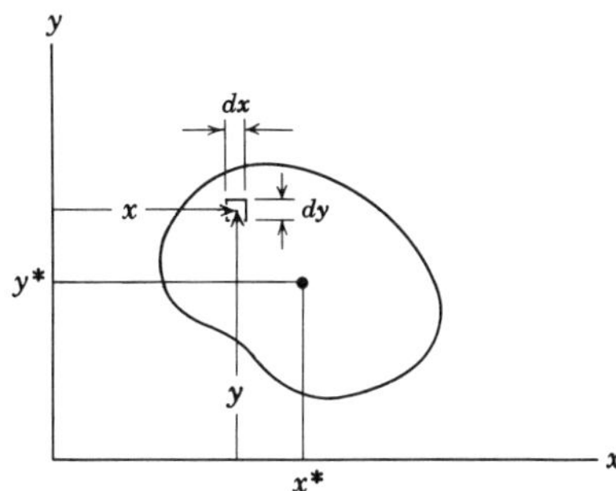

Figure 5.4

$\int_{\text{area}} dA = \int_{\text{area}} dx\,dy$

We have defined certain characteristic points using first moments (moments with distance to the first power). Let us now gain some experience in computing centroids, centers of mass, and centers of gravity.

The first problem is to locate the centroid of the triangular area of Fig. 5.5. Let us compute the area

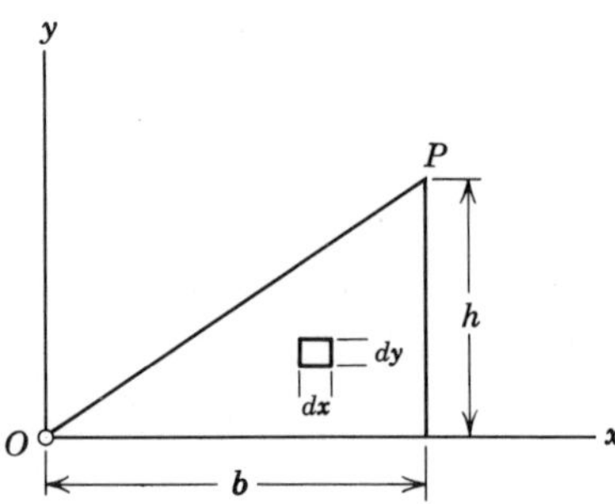

Figure 5.5

of the triangle by multiple integration. Symbolically the area equals $\int_{\text{area}} dA = \int_{\text{area}} dxdy$. We will change this to an iterated integral, integrating first with respect to y. Thus we consider x as constant and integrate along a vertical strip. At constant x, the limits are $y = 0$ and the maximum value of y appropriate to the constant value of x. The equation of the line OP is
h/b $y =$ ___x. Therefore, for a specified x, the upper limit
hx/b of the integration with respect to y is $y =$ ___.

The integration with respect to x proceeds from
b the limit $x = 0$ to the limit $x =$ __.
$\int_{x=0}^{x=b} \left[\int_{y=0}^{y=hx/b} dy \right] dx$ Written in full, the integral for the area is $A =$ _____.

The integration with respect to y is simply the integral of dy. Therefore, between the limits indicated,
hx/b we determine the quantity in brackets to equal ___.

Similarly, integrating hx/b between the limits of $x = 0$ and $x = b$ yields the answer for the area as A
$hb/2$ $=$ ___.

With A known, we can locate the centroid from the formulas $x^*A = \int_{\text{area}} x\, dA$ and $y^*A = \int_{\text{area}} y\, dA$. Seeking x^* first, we write $dA = dxdy$ and write an appropriate iterated integral. This integral is the integral for area, with an additional factor of x in the integrand. Again integrate first with respect to y and write the iterated integral as $\int_{\text{area}} xdA =$ __________.

$$\int_{x=0}^{x=b} \left[\int_{y=0}^{y=hx/b} dy \right] xdx$$

Evaluation of this integral yields $\int_{\text{area}} xdA$ equals ___.

$hb^2/3$

If this first moment of area is divided by the area, A, the x coordinate of the centroid, x^*, is found to equal ___.

$2b/3$

Now we want y^*. For this we use the equation $y^*A =$ ____.

$$\int_{\text{area}} ydA$$

Again we integrate first with respect to y and we write our iterated integral as $\int_{\text{area}} ydA =$ __________.

$$\int_{x=0}^{x=b} \left[\int_{y=0}^{y=hx/b} ydy \right] dx$$

Evaluating this integral and dividing by the area, we find $y^* =$ ___.

$h/3$

For additional practice, you might choose an arbitrarily shaped triangle and show that its centroid is always located at a point 1/3 the distance along a line from the center of each base to the opposite vertex.

Next, we shall determine the centroid of a volume. For a body of uniform density in a constant gravitational field, the centroid is also the center of mass and gravity. The right circular cone in Fig. 5.6

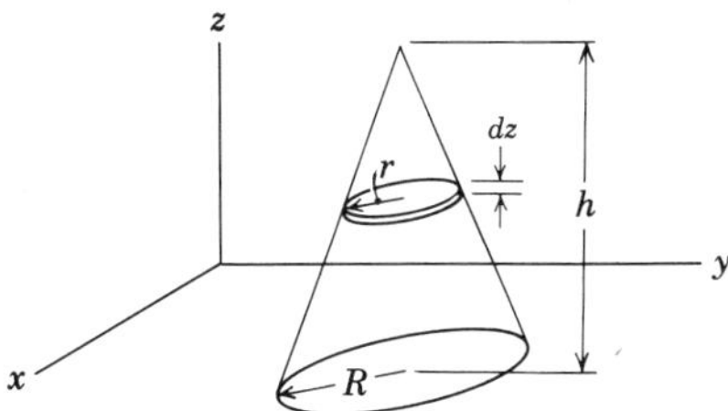

Figure 5.6

rests on the x-y plane and has a height h. Since symmetry shows that the centroid lies on the axis of the cone, we need determine only z^*. To use the formula $z^*V = \int_{\text{vol}} zdV$, we must calculate the volume of the cone. To evaluate dV, in cylindrical coordinates, we choose our volume element to be a thin disk as shown in Fig. 5.6. We can write the expression for dV in terms of the thickness and radius of this element as $dV =$ ____.

$\pi r^2\, dz$

We now have dV in terms of r, but our integration must be with respect to z. To carry out this integration we express r as a function of the variable __.

z

Consider the similar triangles associated with a vertical section of the cone. Geometrically, $r/(h-z) = R/h$. Thus we write $\int_{\text{vol}} dV =$ ________.

$\pi \int_{vol} (h-z)^2 \left(\frac{R}{h}\right)^2 dz$

Since we have in effect performed the integrations with respect to x and y by our choice of the volume element, we now have an integral in terms of the variable z only. What are the limits of integration? ____ to ____.

$z = 0$ $z = h$

Carry out this integration and find that $V =$ ____ for the cone.

$(\pi/3)R^2h$

Now let us evaluate $\int_{\text{vol}} z\, dV$. Again it is desirable to use a disk-shaped element. Again we express dV as $\pi r^2 dz$ and r as a function of z. Having done so, we see that dV is in terms of z only and the integral $\int_{\text{vol}} z dV$ can be written as ________.

$\pi \int_{z=0}^{z=h} z(h-z)^2 \frac{R^2}{h^2}\, dz$

Perform this integration and find that $\int_{\text{vol}} z dV =$ ______.

$(1/12)\pi R^2h^2$

The final step in finding z^* is to divide this value by the volume of the cone. Do this and determine that z^* equals __.

$h/4$

We shall now examine some symmetry properties of some areas and volumes. Figure 5.7 shows an area

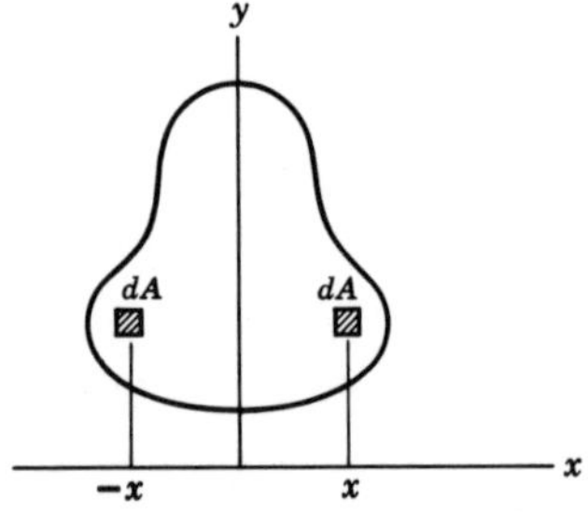

Figure 5.7

that is symmetric about the *y* axis. If we consider the computation of a first moment with respect to this axis, we see that for every element *dA* located at $+x$, there is a corresponding element at $-x$. Thus the first moments of these areas cancel each other. Since this is true for every element of area in the figure, we conclude that the *x* coordinate of the centroid of the area must be ___. zero

This means that the centroid must lie on the __ *y* axis, which is the line of ______. symmetry

For an area with two axes of symmetry, a similar argument shows that the centroid must lie on each of these axes and, therefore, must lie at their _______. intersection

We now apply a similar argument to a three-dimensional body as in Fig. 5.8. The *x-z* plane is a

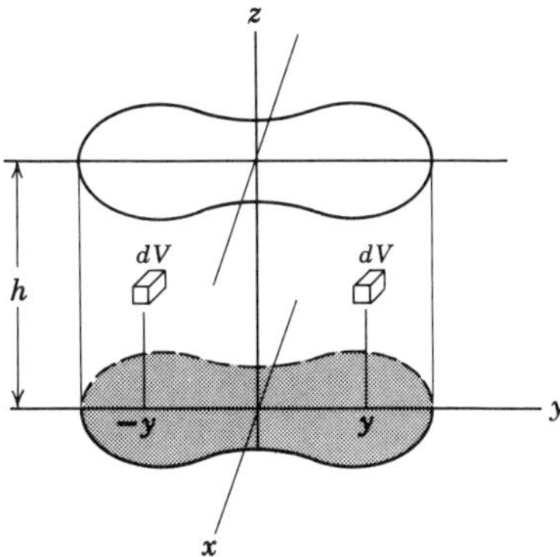

Figure 5.8

plane of symmetry for this body. In the determination of y^*, every element *dV* at a distance $+y$ from the *x-z* plane has a matching element at a distance $-y$ from that plane. We conclude that the *y* coordinate of the centroid of this volume must equal ___ and thus the centroid will lie in the __ plane. zero *x-z*

The *y-z* plane is also a plane of symmetry. We conclude that (__)* equals zero and the centroid lies in *x* the __ plane. *y-z*

Clearly with these two planes of symmetry the centroid must lie somewhere along the __ axis. *z*

In Fig. 5.8 a third plane of symmetry, parallel to

$h/2$ the x-y plane, is located at a distance ___ above the x-y plane. We conclude that the centroid of the object
0 0 shown is at a point with coordinates $x^* =$ __, $y^* =$ __,
$h/2$ and $z^* =$ ___.

We now introduce a new and useful concept. Since many complex areas and bodies consist of a collection of simpler elementary shapes, we present what is called the *composite system approach.* Let us apply this approach to determine the centroid of a plane area. Figure 5.9 shows a triangle from which a

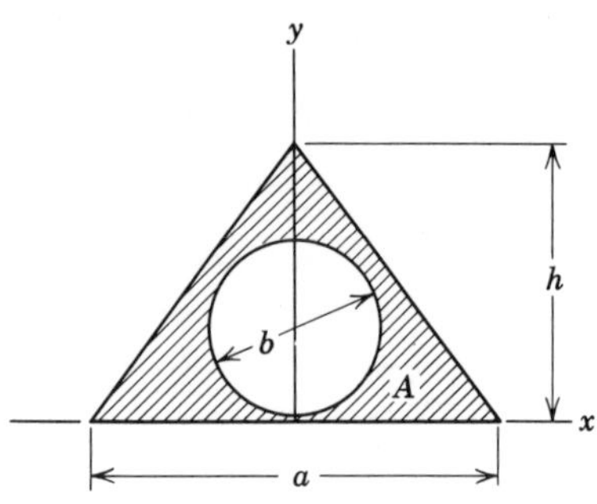

Figure 5.9

circle has been removed. It is obvious that the x coordinate of the centroid equals ___ due to _______.
zero symmetry

To find y^*, we need the total shaded area, A, and its first moment. Direct integration over a region bounded by a circle and a triangle is awkward, and we seek a simpler approach. First, recall the definition $y^*A = \int_{\text{area}} y\, dA$. The simplest way to determine A is to subtract the area of the circle from the area of the
h $b/2$ triangle. The shaded area thus equals $\frac{1}{2}a$ __ $-\pi$ (___)^2.

Now let us consider first moments. We want $\int_{\text{area}} ydA$ for the shaded area. A simple way is to find $\int_{\text{area}} ydA$ for the entire triangle and then to subtract the $\int_{\text{area}} y\, dA$ of the circle. The values of $\int_{\text{area}} ydA$ for the triangle and for the circle are readily determined, since we know the location of the centroid of the triangle and of the circle. For the triangle, y_t^* equals $h/3$ and for the circle,

$y_c{}^*$ equals $b/2$. Therefore, $\int_{triangle} y\,dA = y_t{}^*A = (h/3)$ (___) — $ah/2$

and $\int_{circle} ydA = (b/2)$ (___). — $\pi b^2/4$

We have the area A as the difference between the areas of the triangle and the circle. Similarly we find the first moment of A as the difference between the first moment of the triangle and of the circle. The defining relation for the y coordinate, y^*, of the centroid of A is $y^*A = \int_t ydA - \int_c ydA$. However, we have expressions for A, $\int_t ydA$ and $\int_c ydA$. Thus, $y^*A = y^*\,[(ah/2)$ $-$ ____$] = (h/3) \cdot (ah/2) -$ ________. — $(\pi b^2/4)$ $(b/2) \cdot (\pi b^2/4)$

When numerical values of a, b, and h are given, the centroid of the figure is readily obtained.

The composite approach can also be used to locate centroids of volumes. For example, consider the right circular cylinder-cone combination of Fig. 5.10. Due to symmetry, the centroid must lie somewhere on the __ axis. — z

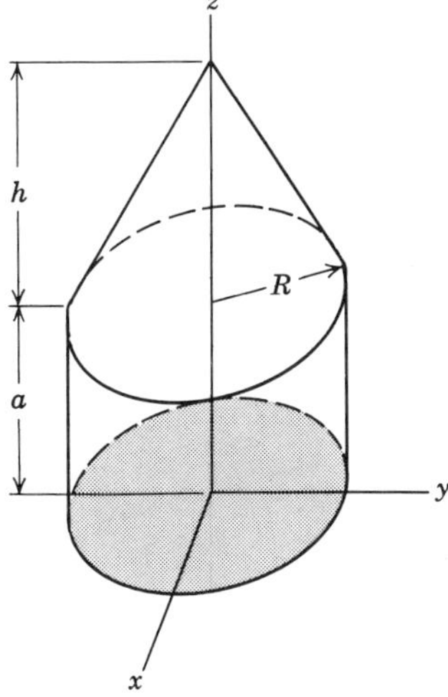

Figure 5.10

We will determine the location, z^*. To begin, we will need the total volume of the cone and cylinder. Then we write $z_t{}^*V = \int_{vol} z\,dV$ for the composite body. We know for the cylinder, that z^* equals ___. Thus, — $a/2$

volume $\int_{cyl} z dV$ is the product of $a/2$ and the _____ of the cylinder.

Likewise, z^*, for the cone in these coordinates is $a + (h/4)$. The $\int_{cone} z dV$ equals $[a + h/4]$ times the volume of the cone. Therefore, for the composite cone-cylinder we have the relation z_t^* (total V) $= \int_{cyl} z dV + \int_{cone} z dV$. Thus $z_t^* (V_{cyl} + V_{cone}) = z^*_{cyl} V_{cyl} + z^*_{cone} V_{cone}$.

Use this relationship to find z_t^* if a, h, and R all
0.688 ft equal one foot. z_t^* equals _____.

This composite approach can be extended in a similar manner to find the centers of mass and centers of gravity of composite bodies.

With this, we end our examination of the foundation of statics. Let's move on to some applications.

Summary

The center of gravity (CG) of a collection of particles is that point through which the resultant of the gravitational forces on the particles acts, regardless of the orientation of the particles relative to the direction of the gravitational attraction. The CG is defined by its coordinates x^*, y^*, and z^* where

$$x^* = \frac{\sum_i x_i w_i}{\sum_i w_i}, \quad y^* = \frac{\sum_i y_i w_i}{\sum_i w_i}, \quad \text{and } z^* = \frac{\sum_i z_i w_i}{\sum_i w_i}$$

Thus the location of the CG is associated with the average first moment of the gravitational forces. For continuous bodies the summations are replaced by volume integrals:

$$x^* = \frac{\int x\rho g\, dV}{\int \rho g\, dV}, \quad y^* = \frac{\int y\rho g\, dV}{\int \rho g\, dV}, \quad \text{and } z^* = \frac{\int z\rho g\, dV}{\int \rho g\, dV}$$

Note that the denominator in each case is simply the total weight of the system.

The related concept of center of mass (CM) is defined without considering any gravitational field. The CM is located by the coordinates x^*, y^*, z^* where

$$(x^*, y^*, z^*) = \frac{\int (x,y,z)\rho\, dV}{\int \rho\, dV}$$

The denominator is the total mass of the system and clearly the CM and CG coincide if g is constant.

Since ρ is often a constant, we frequently are concerned with finding a purely geometrical point, the centroid of a volume, defined by x^*, y^*, z^* where

$$(x^*, y^*, z^*) = \frac{\int (x,y,z)\, dV}{\int dV = V}.$$

Also important is the centroid of an area, x^*, y^*, where

$$(x^*,y^*) = \frac{\int (x,y)\, dA}{\int dA = A}.$$

From these definitions we note that the centroid must lie on any plane or line of symmetry.

Finally, the composite system approach is presented in which bodies are divided into common elementary shapes having known centroid locations and appropriate first moments are obtained by summing the known first moments of the elementary shapes.

Problems

(1) Find the center of gravity of the three particles shown in Fig. P5.1.

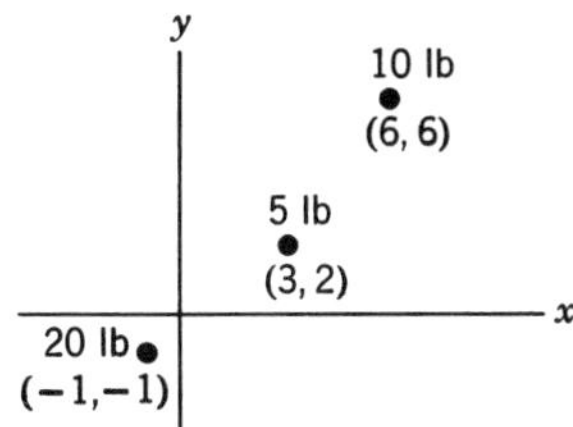

Figure P5.1

(2) Find the centroid of the shaded area in Fig. P5.2.

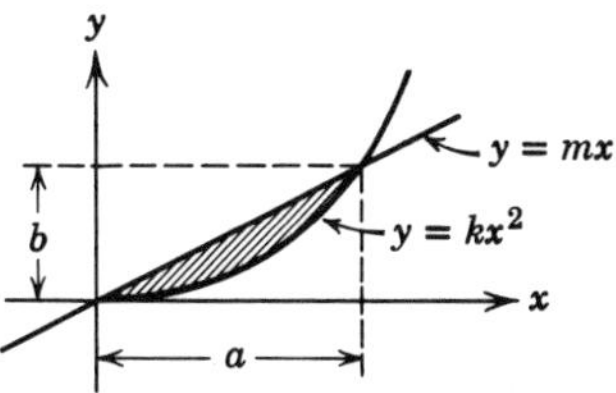

Figure P5.2

(3) In Fig. P5.3, find the centroid of the shaded area. Note the centroid of a semicircular area is $4r/3\pi$ from its straight edge.

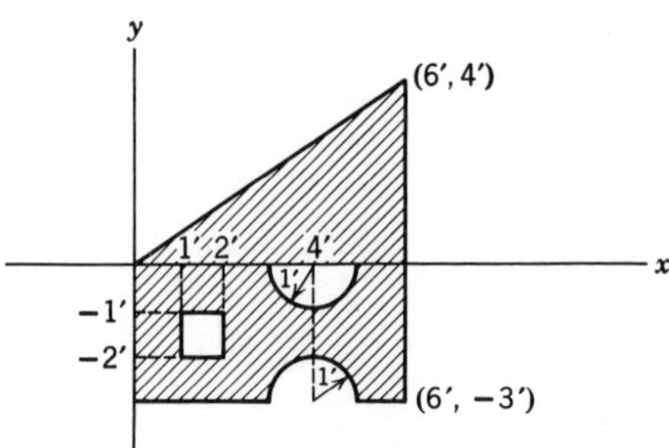

Figure P5.3

(4) Find the resultant force and its location for the distributed load shown in Fig. P5.4.

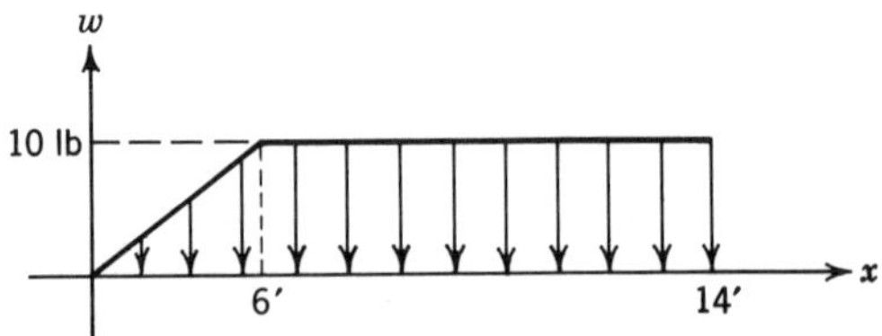

Figure P5.4

(5) Find the center of mass of the homogeneous solid shown in Fig. P5.5.

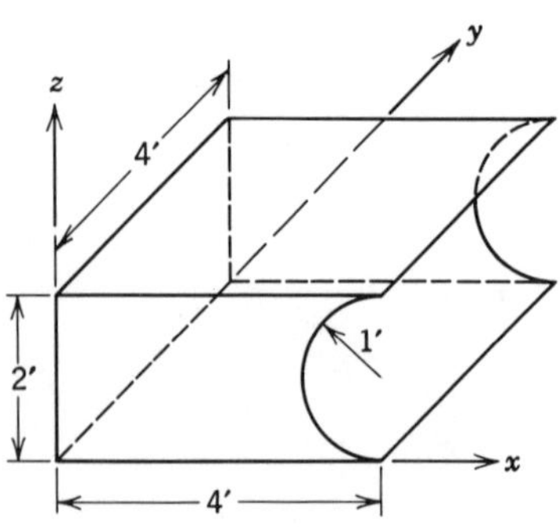

Figure P5.5

(6) Find by integration the center of gravity of the hemispherical shell shown in Fig. P5.6 ($t << 2'$).

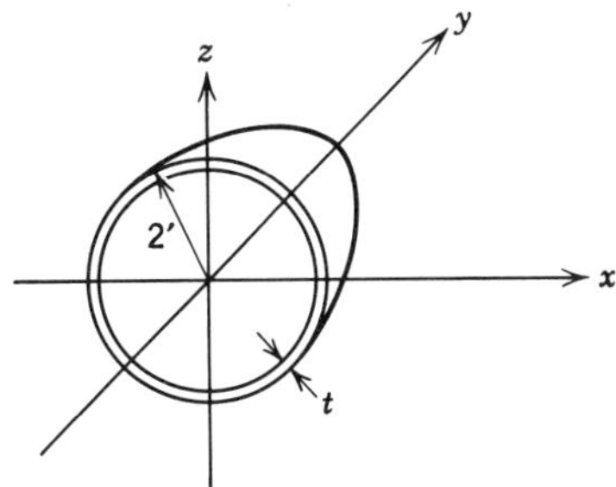

Figure P5.6

Answers to Problems

Chapter 5

(1) $\bar{x} = \frac{11}{7}, \bar{y} = \frac{10}{7}$

(2) $\bar{x} = \frac{a}{2}, \bar{y} = \frac{2b}{5}$

(3) $\bar{x} = 3.4', \bar{y} = -0.2'$

(4) $\bar{R} = -110\bar{j}$ lb
$\bar{x} = 8.36'$

(5) $\bar{x} = 1.6'$
$\bar{y} = 2'$
$\bar{z} = 1'$

(6) $\bar{x} = 0 = \bar{z}$
$\bar{y} = 1'$

chapter 6

Equilibrium

Objectives

Many important engineering systems are in equilibrium. Here we present the general concepts and equations of equilibrium and a general technique for solving problems plus a number of special cases where simplification is possible. Successful completion of this chapter should allow the student to:

1. State the conditions necessary for equilibrium of a particle and a rigid body and write the equations reflecting these conditions.
2. Define a free body diagram and draw correct free body diagrams for simple typical engineering problems.
3. Describe a general procedure for solving equilibrium problems and apply it correctly to solve representative problems of equilibrium of systems of particles and of rigid bodies in two and three dimensions.
4. State the special characteristics of two and three force members and recognize such members in simple problems.

This chapter introduces the central topic of statics, the study of systems in equilibrium. In statics, equilibrium means a condition of rest or motion with *constant* velocity. Thus the identifying characteristic of systems in equilibrium is zero ________.

acceleration

Examples of important engineering systems in equilibrium are numerous. A car jack, a microwave relay tower, and a suspension bridge are examples of systems at rest and therefore in ________.

equilibrium

A less obvious example of a system in equilibrium is a jet airplane flying at constant velocity. The velocity must be constant in magnitude and ______.

direction

Newton's laws state that a particle in equilibrium has zero resultant external force acting on it. In your own words, what does this mean? ______________

This means that the sum of the external forces acting on a particle at rest or in motion with constant velocity is zero (or something similar).

We can use this law to determine if a system is in equilibrium by determining the sum of the forces acting on it. Alternatively, we may determine relationships among the forces acting on a system that is known or assumed to be in equilibrium.

This requirement can be considered in vector or scalar form, since a vector can be represented in terms of scalar components. Thus the equilibrium requirement, namely, that the resultant force on a particle is zero, is equivalent to the scalar requirement that all ________ (number) components of this force equal zero.

three

We conclude that the requirement for equilibrium of a particle in space is expressed as one ______ (type) equation or as three ______ (type) equations.

vector

scalar

As a first example, consider a helicopter undergoing a lift test. The aircraft, moored to the ground by three flexible cables attached at points *A*, *B*, and *C* (Fig. 6.1), exerts a force of $5000\mathbf{j}$ lbs on the hook *H*. We seek the tensions in the cables.

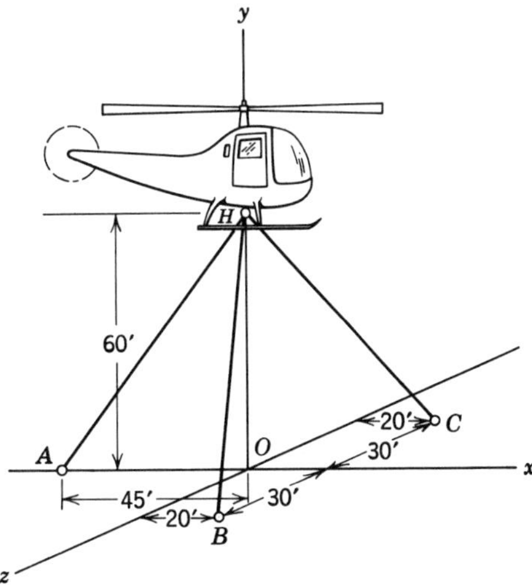

Figure 6.1

The cables exert forces on the hook directed from it through the points *A*, *B*, and *C*. These tensile forces are concurrent at the hook, *H*. Since the lift force, $\overline{\mathbf{L}}$, also acts through *H*, we have a concurrent force system.

We consider the hook *H* as a particle and draw a picture of it (Fig. 6.2). Here we show the hook, cut from the system, and all of the ____ acting on it.

forces

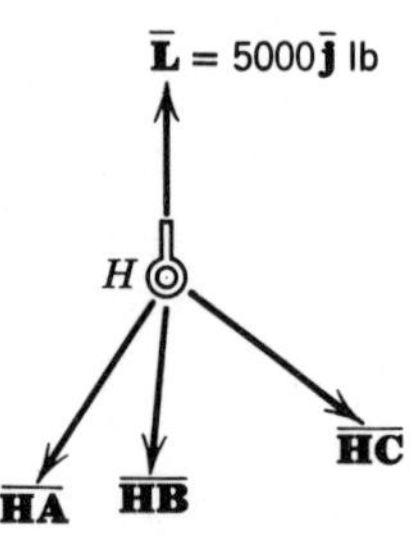

Figure 6.2

These forces are vector quantities of known direction and unknown ________.

magnitude

We now express each of these forces as the product of a known unit vector and an unknown magnitude. Consider the force $\overline{\mathbf{HA}}$. The triangle *OHA* of Fig. 6.1 is a three-four-five triangle. Therefore, a unit vector from *H* through *A* is $\bar{\mathbf{e}}_{HA}$ = ____$\bar{\mathbf{i}}$ + ____$\bar{\mathbf{j}}$ + ___$\bar{\mathbf{k}}$.

−3/5 −4/5 0

The force $\overline{\mathbf{HA}}$ is written as $(A)\,\bar{\mathbf{e}}_{HA}$, where *A* is the magnitude of $\overline{\mathbf{HA}}$. Next consider the force $\overline{\mathbf{HB}}$. The unit vector in the *HB* direction is found by using ratios of the length of the diagonal to the lengths of the sides of the 20′ by 30′ by 60′ parallelepiped associated with the line *HB*. Compute the distance from *H* to *B*. It is ___ feet.

70

Thus the *x* component of the unit vector $\bar{\mathbf{e}}_{HB}$ is 2/7, the *y* component is −6/7, and the *z* component equals ____.

3/7

The force $\overline{\mathbf{HB}}$ equals the magnitude *B* multiplied by the unit vector $\bar{\mathbf{e}}_{HB}$ = ____$\bar{\mathbf{i}}$ $\bar{\mathbf{j}}$ ____$\bar{\mathbf{k}}$.

2/7 −6/7 3/7

$\overline{\mathbf{HC}}$ equals a magnitude *C* multiplied by a unit vector $\bar{\mathbf{e}}_{HC} = 2/7\bar{\mathbf{i}}$ + ____$\bar{\mathbf{j}}$ + ____$\bar{\mathbf{k}}$.

−6/7 −3/7

We are treating the hook as a particle, and we have written our unknown forces in an appropriate vector form. We can now apply the equations of equilibrium for a particle. The vector sum of the ____ acting on H must equal zero. forces

If we write $\overline{\mathbf{HA}} + \overline{\mathbf{HB}} + \overline{\mathbf{HC}} + \overline{\mathbf{L}} = 0$ and then convert this vector equation into three scalar equations, we can sum the components of the forces in each of the coordinate directions and set these sums each equal to ___. zero

For the x direction, the scalar equation is ___A + ___B + ___$C = 0$. −3/5 +2/7 +2/7

Similarly for the y direction, we obtain the equation ______________. $-(4/5)A - (6/7)B - (6/7)C + 5000 = 0$

For the z direction $(3/7)B - (3/7)C = 0$. On scratch paper, solve these three scalar equations to find $A =$ ____, $B =$ ____, and $C =$ ____. 1920 lb 2020 lb 2020 lb

We now investigate the requirements for equilibrium of a rigid body. For a rigid body to be in equilibrium, the *resultant of its external forces* and the *resultant moment of these external forces* about *any* point must equal zero. A derivation of this statement can be found in most standard texts. Thus, for a rigid body to be in equilibrium, the resultant ___ acting on it and the resultant ____ of the external forces about any point must equal zero. force moment

Rigid body equilibrium requires that two vector quantities, the resultant force and the resultant moment, be zero. Thus the three scalar components of the resultant force and the three scalar components of the resultant moment about any point are zero.

In the general three-dimensional case of equilibrium of a rigid body, we can write a total of ____ (number) scalar equations of equilibrium. 6

Consequently, the equilibrium equations alone can be used to solve for, at most, __ unknown force or moment components for a rigid body. 6

A common and important case of equilibrium is the two-dimensional or plane case in which all forces and couples lie in a single plane. With two coordinate dimensions involved, only two scalar components of any force and two force equilibrium equations will be present. The moments of all forces can be expressed as vectors perpendicular to the plane. The direction of the resultant moment on a rigid body is known. Moment equilibrium for the plane case requires that the magnitude of this moment be zero and thus involves how many scalar equations? _____.

1 (one)

Therefore, for the plane case we have a total of how many independent equations of equilibrium? _____.

3 (three)

Next we present a technique for isolating a particle or rigid body in order to apply the equations of equilibrium. This technique involves drawing *free body diagrams.* Here the word *free* means separated from other parts of the system that are not shown. The process of constructing a free body diagram is simple but important. We begin by choosing a closed surface containing some part of a system. We then draw a diagram of the part alone. Next we add to the diagram all the forces on the body that act *across the surface.* The force of gravity, for example, should be shown, since it acts across the boundary. The force of one internal part on another should not be shown. Why? ______________________________.

Such forces do not act across the boundary

Often free body diagrams will include unknown *reaction forces*, that is, forces at supports. These depend on the external loads applied to the system. It is unnecessary to know the directions of the unknown reactions. Arbitrarily choose one direction as positive and draw your vector representation accordingly. When the magnitude of this reaction is found to be negative, its direction is ____________________ the
the same as/opposite to
direction chosen.

opposite to

Consider the wrecker, in Fig. 6.3, that is lifting a sports car from a ravine at a constant speed. Due to symmetry, the problem is treated as plane. We ask: (1) What is the tension in the cable supporting the boom (*AB*)? (2) What is the load on the pin, *C*? Let's isolate some part of the system so that we can apply the equations of _________.

equilibrium

Since the sports car is being raised at constant speed, equilibrium of the car requires that the tension in the cable *DS* is ____ lb.

2000

We seek information that is related to the equilibrium of the boom *BCD*. We shall isolate the boom by drawing a ____ ____ ______.

free body diagram

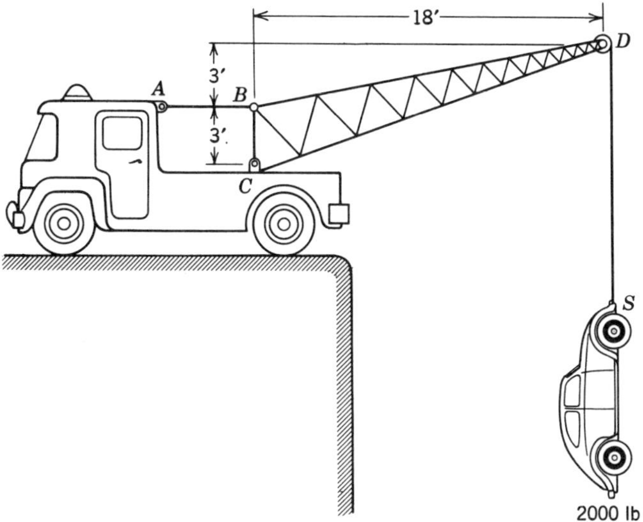

Figure 6.3

Picture a closed surface surrounding the boom *BCD*. In a plane, a free body is isolated by a closed curve. Draw a diagram of the boom in the space provided. Then indicate on this diagram the forces that act ____________ the enclosing curve.
across/within

across

Each of the cables *AB* and *DS* transmits a tensile force that can be represented by a vector of known direction. The pin at *C* in this plane problem has a force of unknown magnitude and direction acting on it. This force may be represented by ________
(number)
scalar components.

two

Add these forces to your diagram and compare it with Fig. 6.4. In this free body diagram there are

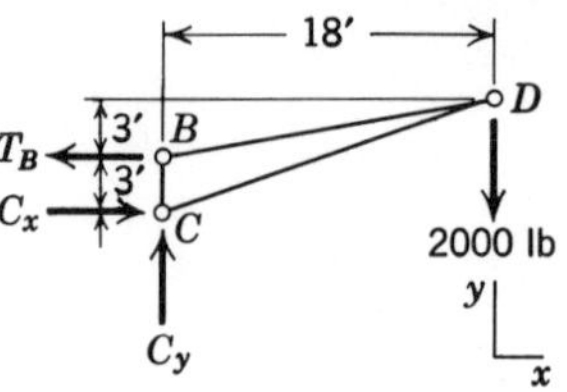

Figure 6.4

three unknown quantities. We now write the equilibrium equations for the boom. Two scalar force equations, in the x and y directions, and one moment equation will be used. For the moment equation, it is desirable to sum moments about a point that has a maximum number of unknown force components passing through it. This leads to equations with a minimum number of unknowns present. Logical choices of points about which to take moments would thus be ____.

B or *C*

Let us now write the three equations of equilibrium. First, for equilibrium of x components, we write __ − __ = 0.

C_x T_B

Summing force components in the y direction leads to the equation ______ = 0.

$C_y - 2000$

Finally, we write a moment equation about C. Since this is a plane problem, we need write only a scalar moment equation. Calling the counterclockwise direction positive, our equation $M_C = 0$ reads ____________.

$3T_B - (18) \cdot (2000) = 0$

Solve these three equations for the unknowns T_B, C_x, and C_y. Find that $T_B =$ ______, $C_x =$ ______, and $C_y =$ _____.

12,000 lb. 12,000 lb.
2000 lb.

You have almost finished the problem. However, don't forget to check your answers. A good check is to take moments about some other point, say D. Do this and verify that the sum of the moments about point D equals zero.

We now solve a three-dimensional problem. Consider the L-shaped radar antenna structure shown in equilibrium in Fig. 6.5. We are asked to find the two

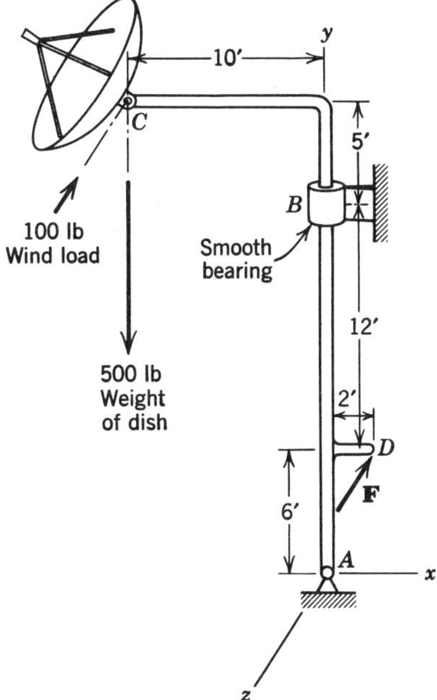

Figure 6.5

components of the force acting at the bearing *B*, and the magnitude of the force $\bar{\mathbf{F}}$ acting on the positioning lever at *D*.

In order to write equations of equilibrium, we isolate the system by drawing a free body diagram. Draw a free body diagram of the L-shaped member in the space provided and then compare it with Fig. 6.6.

We can classify the force system involved as a general _____ dimensional system. three-

Examination of the free body diagram shows a

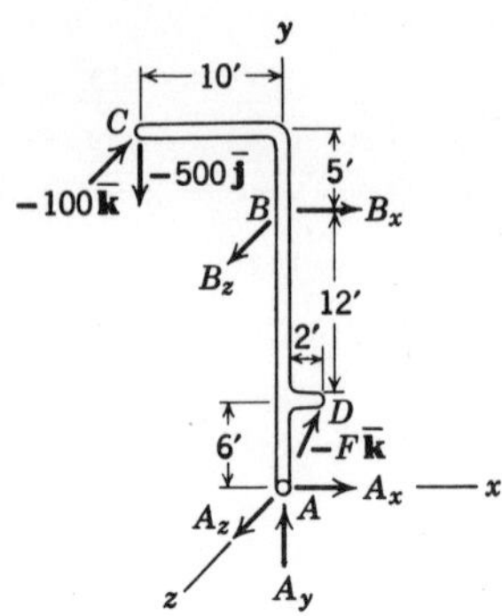

Figure 6.6

six (A_x, A_y, A_z, B_x, B_z, F) total of ____________ unknown scalar quantities
(number)
present.

Since we can write six independent equations of equilibrium for this free body, we can solve for the unknowns present. However, we have been asked to find only B_x, B_z, and F. We would like to determine these unknown quantities without solving for the reaction components at A. We can do this by writing equations of moments about axes which pass through point A. If an axis passes through A, then the reaction
zero force at A will have ___ moment about this axis.

First, we determine the magnitude of $\bar{\mathbf{F}}$. The simplest way is to write one equation in the one unknown F. To obtain this equation, sum the moments about
B a line through points A and __.

We could determine moments about the line by taking moments of the forces about some point on the line and then taking their scalar product with a unit vector parallel to the line. However, the force system is sufficiently simple so that we can use a direct scalar approach.

The moment of force $\bar{\mathbf{F}}$ about line AB is $2F$. The
0 moment of the weight force, $-500\bar{\mathbf{j}}$ lb., is __ and the
−1000 moment of the wind load is ____ ft-lb.

Moment equilibrium about line AB is thus ex-
1000 pressed by the equation $2F -$ ___ $= 0$.

500 From this equation, we find $F =$ ___ lb. Now, how can we find B_x? A single equation with B_x as the only unknown can be written. Since B_z and F are parallel,

the equation sought is a moment equation about the __ axis which passes through point __. z A

From the free body diagram, we see that B_z and the wind force are parallel to the z axis. Thus, neither contributes a moment about the z axis. Again we can write a simple scalar moment equation, (500) · (10 ft) $- B_x \cdot$ (___) = 0 and solve for B_x = _____. 18 ft 278 lb

Finally, we solve for B_z. This can be accomplished by writing a moment equation about a line through point *A* parallel to the __ axis. x

The component B_x does not contribute to this moment since it is parallel to the axis. The weight (force) of the dish does not contribute because it ________ the x axis. intersects

We thus have the moment equation 18 B_z − 6 · (________) − (___) · (_____) = 0. *F* = 500 lb 23 ft 100 lb

From this equation we find B_z = _____. 294 lb

We now have B_x, B_z, and *F*, the unknown quantities sought. To check the values of B_x, B_z, and *F*, sum moments about point *A*. If you wish, consider force equilibrium and find the components A_x, A_y, and A_z. You should find that A_x = ______, A_y = _____, A_z = _____. −278 lb 500 lb 306 lb

Let us review the procedure we have used for the solution of equilibrium problems. First, determine what information is given and what results are to be obtained. Next, isolate a certain body or group of bodies associated with the desired unknown forces by drawing a ______________. free body diagram

Now classify the problem according to the type of force system present. Then determine the number of independent equations of equilibrium available and compare this number with the number of unknown quantities to be determined. If there are enough independent equations to solve for the unknowns, do so. If more independent equations are required, they can be obtained by drawing free body diagrams of other parts of the system. This process is repeated until we are able to solve for the unknowns present or have used all available free body diagrams. In this latter case, we cannot solve our problem using statics alone and the problem is called *statically indeterminate.*

It is tempting to generate additional equations

by writing moment equations about additional points. However, these additional equations are not *independent* and, therefore, cannot be used to solve for additional unknowns.

We shall now examine two special and common cases of rigid body equilibrium. The first is a two-force member, a rigid body in equilibrium under the action of *two* forces. In Fig. 6.7, a force $\bar{\mathbf{F}}$ is applied at P. For equilibrium the other force must equal ___.

$-\bar{\mathbf{F}}$

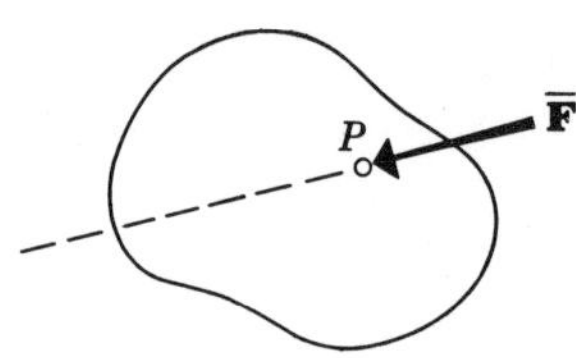

Figure 6.7

Where must this force be located? See if you can determine this on your own before proceeding.

In general the force $\bar{\mathbf{F}}$ and the force $-\bar{\mathbf{F}}$ form a couple. Since these are the only forces acting on the body, the magnitude of this couple must be ___.

zero

The only way to insure that the couple is zero is for the forces $\bar{\mathbf{F}}$ and $-\bar{\mathbf{F}}$ to have the same ______.

line of action

Thus, in order for a body to be in equilibrium under the action of two forces, the forces must be ___ in magnitude, ___ in direction and have the ___ line of action.

equal
opposite same

For a body to be in equilibrium under the action of two forces, one at P and the other at Q (Fig. 6.8.), these forces must act along the line ___ regardless of the shape of the body.

PQ

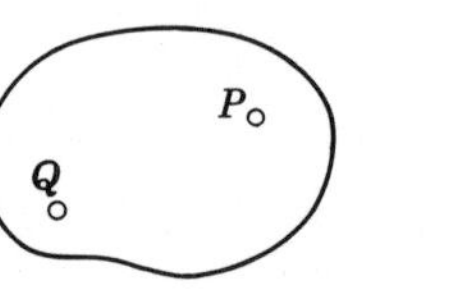

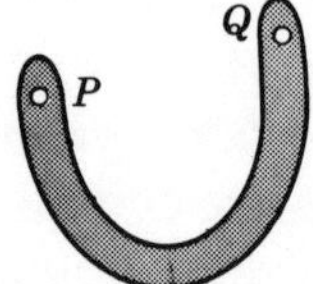

Figure 6.8

The member shown in Fig. 6.9 ________ be
can/cannot
maintained in equilibrium by a single force at *Q*.

cannot

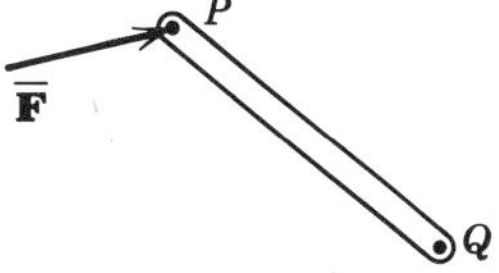

Figure 6.9

The other special case is the equilibrium of a body acted on by three forces, Fig. 6.10. For the body to be in equilibrium, the sum of the forces must equal ___.

zero

Consider the case in which two of the forces $\overline{\mathbf{F}}_1$ and $\overline{\mathbf{F}}_2$ intersect. We can replace these two forces by their resultant acting through the point of ________.

intersection

We now have a body in equilibrium under the action of two forces. Therefore, $\overline{\mathbf{F}}_3$ must have the same line of action as the resultant $\overline{\mathbf{F}}_1$ and $\overline{\mathbf{F}}_2$, and the three forces are ________.

concurrent

Furthermore, since $\overline{\mathbf{F}}_1$ and $\overline{\mathbf{F}}_2$ and their resultant

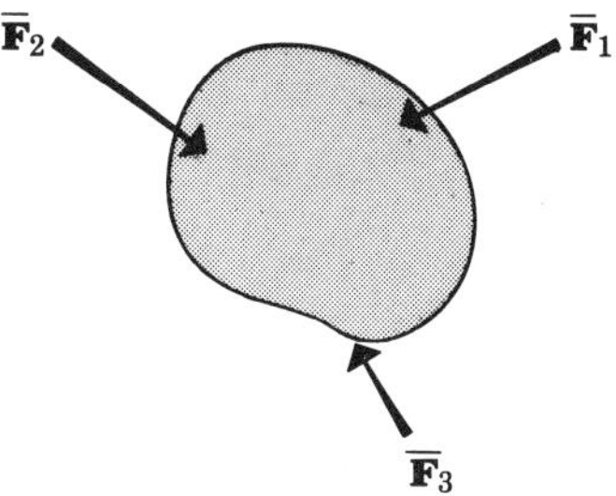

Figure 6.10

lie in the same plane, $\overline{\mathbf{F}}_1$, $\overline{\mathbf{F}}_2$, and $\overline{\mathbf{F}}_3$ must lie in the same ___.

plane

Consider now that no two forces intersect. For the resultant moment about *any* point on $\overline{\mathbf{F}}_3$ to equal zero, the moments of $\overline{\mathbf{F}}_1$ and $\overline{\mathbf{F}}_2$ about that point must be equal and opposite. Draw some diagrams to convince yourself that this is possible only if $\overline{\mathbf{F}}_1$, $\overline{\mathbf{F}}_2$, and $\overline{\mathbf{F}}_3$

lie in the same plane. Since $\bar{\mathbf{F}}_1$, $\bar{\mathbf{F}}_2$, and $\bar{\mathbf{F}}_3$ are coplanar and none intersect, they must be ______.

parallel

The spacing of the three forces must be such that their resultant force and moment equals ___.

zero

In summary, if a body is in equilibrium under the action of three forces, then the three forces must either be coplanar and concurrent or coplanar and ______.

parallel

We now consider composite systems, that is, systems consisting of a number of components. Figure 6.11 shows a plane system consisting of two

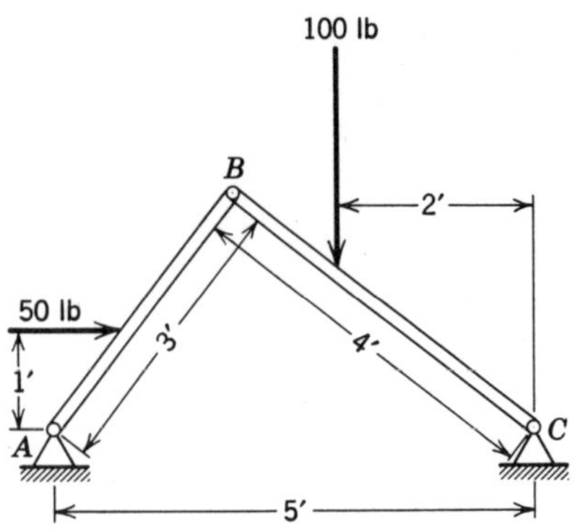

Figure 6.11

straight members, *AB* and *BC*, connected at their ends by pins. We seek the reaction forces at *A* and *C*, and the force acting at the pin *B*. Draw a free body diagram in the space provided and compare with Fig. 6.12.

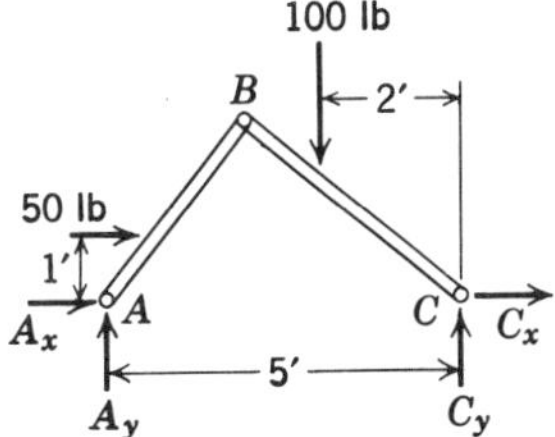

Figure 6.12

We now write equations of equilibrium. Since this is a plane problem, we can write at most ______ (number) independent equations. — three

Figure 6.12, shows ______ (number) unknown quantities. — four

Therefore, we ______ (can/cannot) solve for *all* of these unknowns without additional equations. — cannot

However, let us solve for as many as we can. We can solve for A_y and C_y by writing moment equations. A moment equation about point A contains only the unknown C_y and can be used to determine C_y. Similarly, if we take moments about point __, the only unknown force component present is __, and we can solve for it. — C; A_y

Perform these operations. The equation for the sum of the moments about A is ______. — $-50 - 300 + 5C_y = 0$

Solve this equation and find that $C_y =$ ____. — 70 lb

Next, write the equation for the moments about point C: ______. — $-5A_y - 50 + 200 = 0$

Thus, A_y equals ____. — 30 lb

There is no equation (based on the free body of Fig. 6.12) that can be solved for either A_x or C_x. However, we can sum forces in the x direction as ______. — $A_x + C_x + 50 = 0$

Thus we can determine the sum of A_x and C_x, but we cannot determine their values. For the values, we must obtain additional independent equations through use of additional free body diagrams. Where should we divide the system? ______ Why? ______ — Divide it at point B, since we have information about the forces transmitted from one member to the other there.

Since the joint is pinned, there is only a

force

moment

_____________ transmitted at B, and the _____
force/moment force/
_______ transmitted is zero.
moment

Next, separate the two members at B and draw a free body diagram of AB.

The point A is the same as in Fig. 6.12. At point B we must show the force transmitted from member BC. We represent the force on B in terms of its rectangular components B_x (→) and B_y (↑).

To draw a diagram of member BC, we must decide how to draw the forces at B. We obey Newton's third law; for every action there is an equal and opposite reaction. Therefore, if member BC exerts the forces just drawn on member AB at B, then AB must exert equal and opposite forces on member BC at B. Since we have drawn B_x as positive to the right on member AB, it must be shown as positive to the

left

________ on member BC. B_y is positive upward on
right/left

downward

AB; therefore it must be shown as positive _______
upward/
_________ on member BC.
downward

Draw a free body diagram of member BC following the rules just established. Check your diagrams with Fig. 6.13.

We now consider equilibrium of members AB and BC. Since we can write three equations for each, we can solve for the six unknowns.

We now can complete the solution. Consider member AB first. Since we know A_y (30 lb), we can find B_y by using the sum of the forces in the vertical direction. We can solve for A_x and B_x by using moments. We solve for A_x by summing moments about

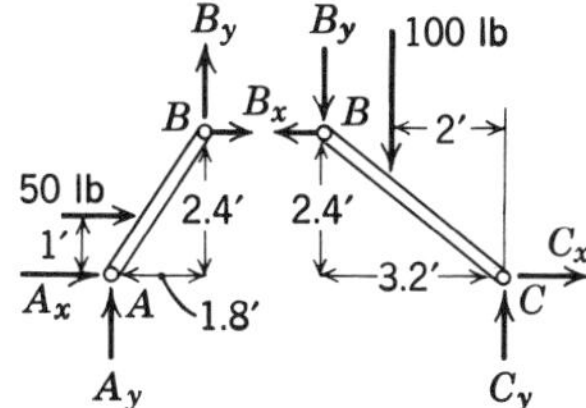

Figure 6.13

point B. Write this moment equation. ______________ ______________.

$2.4A_x - (1.8) \cdot (30) + (1.4) \cdot (50) = 0$

Solve this equation to determine that A_x equals ________.

-6.67 lb

Use force equilibrium for member AB to determine that $B_x =$ ________ and $B_y =$ ______

-43.3 lb -30 lb

Only C_x remains unknown. Force equilibrium of member BC leads immediately to the result that $C_x =$ ________.

-43.3 lb

We can now use additional moment equations as checks.

Since B_x and B_y were found to be negative, the directions chosen for them were incorrect. Thus, the *actual* component directions are opposite to the directions assumed. Do not waste time deciding which way unknown components act. Simply pick a direction; if the answer is negative, recognize that the component acts in the ______________ direction.
same/opposite

opposite

Remember Newton's third law requires that if we assume B_x and B_y positive in one direction on member AB, then we must consider them positive in the ______________ direction on member *BC*.
same/opposite

opposite

Another application of equilibrium involves the determination of forces acting *within* continuous members. In Fig. 6.14 a continuous beam is fixed at B and loaded by a force and moment at A. What are the components of the force and moment transmitted across the section *S-S*? Imagine cutting off the beam at *S-S*. What force and moment at the section are necessary for equilibrium of the right segment? We proceed by drawing a free body diagram, Fig. 6.15.

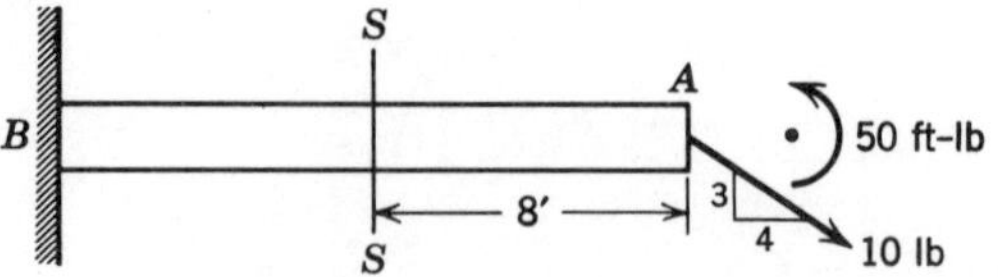

Figure 6.14

three

For this free body we have ___ unknown quantities. Since it is a plane problem, we may solve it readily. Using scratch paper, write and solve appro-

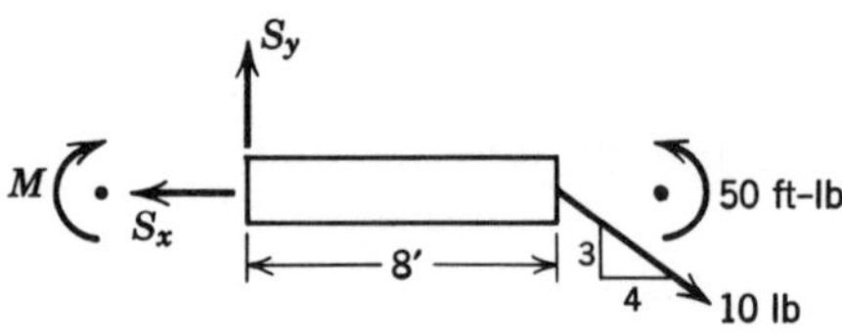

Figure 6.15

8lb 6lb 2ft-lb

priate equations for the three unknowns M, S_x, and S_y. We find that S_x = ___, S_y = ___, M = ___.

Thus by considering the equilibrium of an element imagined cut from the beam, we can determine the resultant force and moment acting at the section.

Summary

In statics, equilibrium means the absence of acceleration, that is, a state of rest or motion with constant velocity. A particle in equilibrium has zero resultant force acting on it, that is, $\overline{\mathbf{F}}_R = 0$. For a rigid body to be in equilibrium, the resultant force acting on it must be zero as must the resultant moment about any point. Thus, for a rigid body $\overline{\mathbf{F}}_R = 0$ and $\overline{\mathbf{M}}_{R-O} = O$ for every point O. Since one vector equation is equivalent to three scalar equations, a maximum of three independent scalar equations of equilibrium can be written for a particle and six for a rigid body.

To solve problems we first isolate an appropriate part of our system by drawing a free body diagram, a diagram of the contents of an imaginary closed surface in space, plus all forces acting across the surface on the contents. Suitable equations of equilibrium are then written and solved if possible. If an inadequate number of equations are available, additional ones are sought by drawing free body diagrams of other parts of the system or by subdividing the initial diagram. This process is continued until the problem is solved or indentified as statically indeterminate.

Bodies in equilibrium under the action of two or three forces are important special cases. For the two-force body, the forces must be equal, opposite, and colinear; for the three-force body, concurrent and coplanar.

The analysis of systems consisting of a collection of components usually requires writing equations for free bodies of individual components. In doing this, the system ordinarily is separated at points where special information on force and moment being transmitted is available (zero moment at a hinge, for example). Care must be taken to apply the action-reaction law so that signs of force components at a separation point are consistently treated.

Problems

(1) Consider a particle P acted on by the known forces $\bar{\mathbf{F}}_1 = \bar{\mathbf{i}} + 2\bar{\mathbf{j}} + 3\bar{\mathbf{k}}$, and $\bar{\mathbf{F}}_2 = -2\bar{\mathbf{i}} + 3\bar{\mathbf{j}} + 4\bar{\mathbf{k}}$, and by the unknown force $\bar{\mathbf{F}}$. If the particle is in static equilibrium, the unknown force $\bar{\mathbf{F}}$ can be determined from the equilibrium conditions. What is $\bar{\mathbf{F}}$?

(2) Consider the very thin rigid body in Fig. P6.1 which is acted

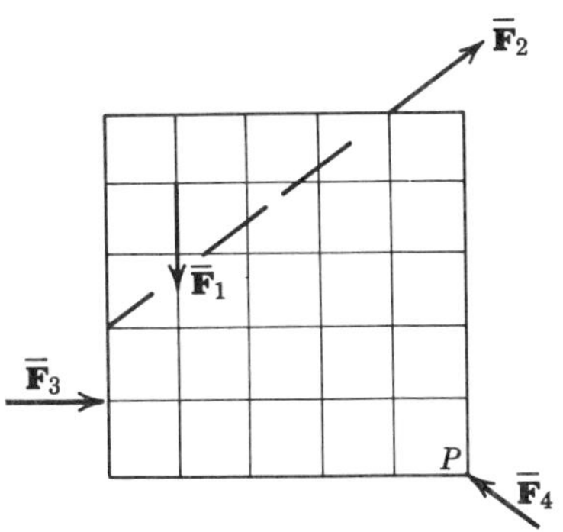

Figure P6.1

upon by the forces $\bar{\mathbf{F}}_1 = 10\bar{\mathbf{i}}$ and $\bar{\mathbf{F}}_2 = -30\bar{\mathbf{i}} + 4\bar{\mathbf{j}}$, the force $\bar{\mathbf{F}}_3$ whose magnitude is unknown and the force $\bar{\mathbf{F}}_4$ which acts through the point P but is otherwise unknown. Determine F_3 and $\bar{\mathbf{F}}_4$.

(3) Consider the body, in Fig. P6.2, and the three points, A, B, and C

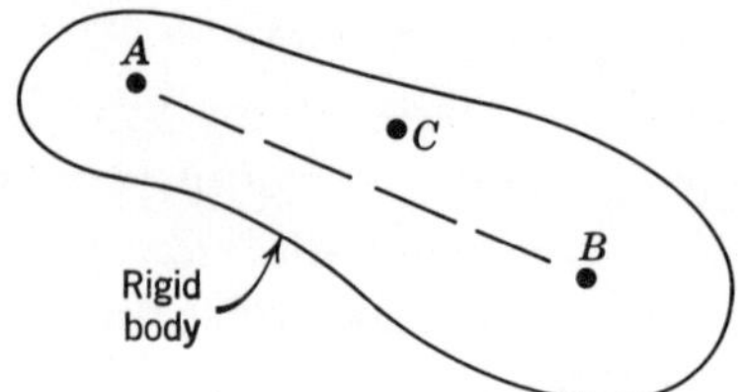

Figure P6.2

in the body. (a) Suppose two forces are applied to the body —one at A and the other at B. If the body is in equilibrium, what can be said about the descriptions of the forces? (b) Suppose three forces are applied to the body—one at A and two at B. The body is in static equilibrium. Describe the forces as completely as possible. (c) Suppose three forces are applied to the body—one at each of the points A, B, and C. Furthermore, suppose the forces are coplanar and one is completely known. Is it possible to completely determine the other two forces from the relations of statics?

(4) Sketch a free body diagram and solve for the unknown reaction forces $\bar{\mathbf{R}}_A$ and $\bar{\mathbf{R}}_B$ in Fig. P6.3.

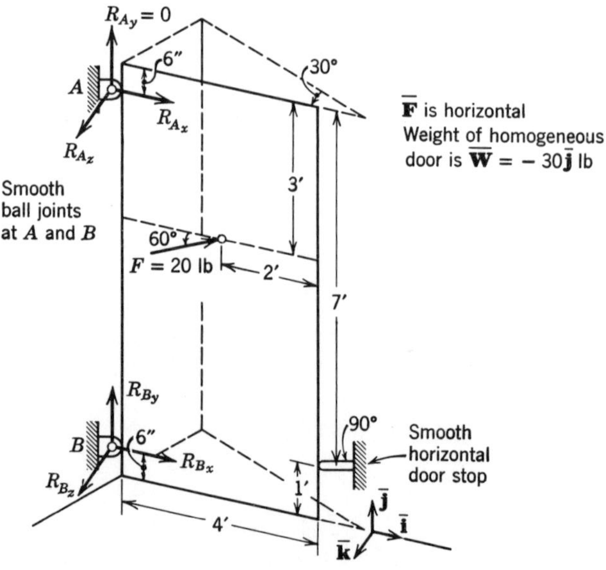

Figure P6.3

(5) Sketch a free body diagram and solve for the unknown reaction force and moment at the wall in Fig. P6.4.

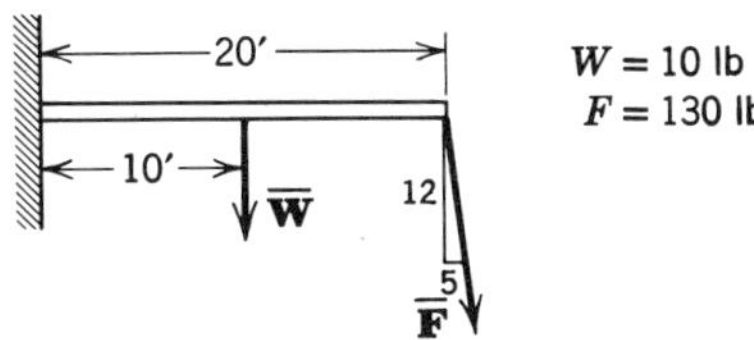

Figure P6.4

(6) Determine the reaction at pin joint *A* in Fig. P6.5. The wall and all pins are smooth. The wheel at the right is in equilibrium under the action of the frame on it at the pin, an external couple as shown, the force $\overline{\mathbf{F}}$ which is just large enough for equilibrium, and the normal reaction due to the surface that the wheel rests on. $\overline{\mathbf{W}}$ is applied to frame section *AC* at point *A*.

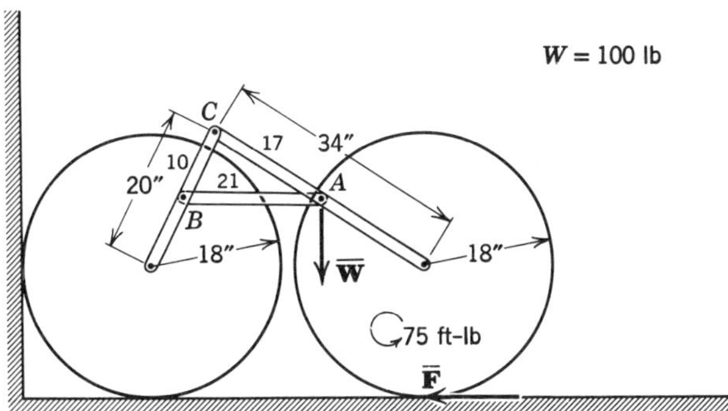

Figure P6.5

Answers to Problems

Chapter 6

(1) $\overline{\mathbf{F}} = \overline{\mathbf{i}} - \overline{\mathbf{j}} - 4\overline{\mathbf{k}}$

(2) $F_3 = 17$

$\overline{\mathbf{F}}_4 = -7\overline{\mathbf{i}} - 21\overline{\mathbf{j}}$

(3) a. $\overline{\mathbf{F}}_A = -\overline{\mathbf{F}}_B$

b. $\overline{\mathbf{F}}_{1B} + \overline{\mathbf{F}}_{2B} = \overline{\mathbf{F}}_B$

c. 3 equations 4 unknowns ∴ in general statically indeterminate

(4) $\bar{\mathbf{F}}_A = 1/14(135\bar{\mathbf{i}} + 420\bar{\mathbf{j}} - 15\sqrt{3}\,\bar{\mathbf{k}})$

$\bar{\mathbf{F}}_B = 1/14(-205\bar{\mathbf{i}} + 85\sqrt{3}\,\bar{\mathbf{k}})$

$\bar{\mathbf{F}}_S = 10\bar{\mathbf{k}}$

(5) $\bar{\mathbf{R}} = -30\bar{\mathbf{i}} + 130\bar{\mathbf{j}} - 40\bar{\mathbf{k}}$

$\bar{\mathbf{M}} = 800\mathrm{j} + 2500\bar{\mathbf{k}}$

(6) $\bar{\mathbf{A}} = \dfrac{475\bar{\mathrm{i}}}{14}$

chapter 7

Introduction to Structural Analysis

Objectives

This chapter will introduce the application of equilibrium principles to the analysis of four important types of engineering structural elements, trusses, frames, beams and cables. Mastery of the material of this chapter should allow the student to:

1. Define and recognize a truss, frame, beam, and cable.
2. Determine desired forces in a simple truss using an appropriate combination of the methods of joints and of sections.
3. Determine unknown force components in simple frames.
4. Define and describe the sign conventions associated with lateral load, shear, and moment in a beam.
5. Derive and explain the differential relations between load, shear, and moment in straight beams.
6. Utilize these relations to construct shear and moment diagrams for beams loaded laterally with both concentrated and distributed forces.
7. Define a flexible cable and recognize a properly posed problem of cable equilibrium.
8. Solve for the unknown quantities in properly posed problems involving cables with concentrated loads.
9. Derive the equations describing plane cables with distributed loads and utilize them to solve simple problems with loads distributed uniformly in the horizontal direction.

7.1 TRUSSES AND FRAMES

We are now ready to apply the principles of statics to practical structures designed to support loads. We begin with *trusses*, which are defined as structures composed of two-force members fastened together to form a rigid system. Here rigid means not only that the individual members do not deform but also that the structure does not collapse. For example, Fig. 7.1(*a*) shows a rigid structure. It is able to withstand the force $\bar{\mathbf{F}}$ without collapsing. Fig. 7.1(*b*) shows a structure for which even a small force $\bar{\mathbf{F}}$ ______ will/will not cause collapse of the system.

will

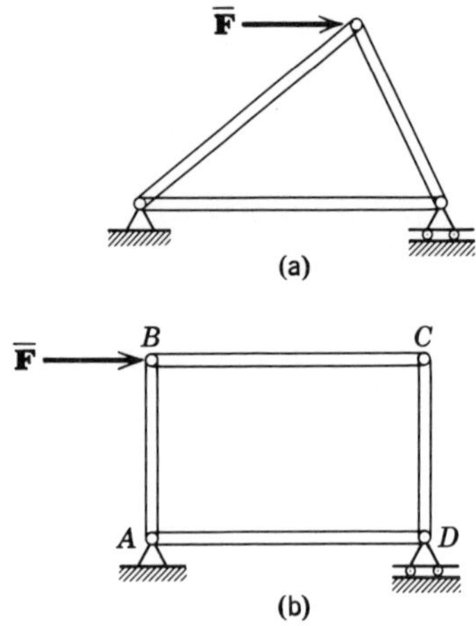

Figure 7.1

The analysis of trusses is based upon the following premises and assumptions: (1) Only two-force members joined by smooth pins are present. (2) All forces act at the pins. (3) The weights of the individual members can be neglected or included as forces acting at the pins.

Trusses, which are common engineering structures, are frequently so complex that they cannot be analyzed by statics alone. However, statics is sufficient to analyze an important class of trusses called *simple trusses.* We will investigate first *plane simple trusses*, that is, trusses lying entirely in one plane.

Let us consider some of the characteristics of the plane simple truss. The arrangement of members in Fig. 7.1(*b*) is not a truss because it will ______ when a force is applied.

collapse

Clearly, this structure can be made rigid by the addition of a member from *A* to *C* or one from __ to __.

B *D*

The structure of Fig. 7.2 is rigid. However, note

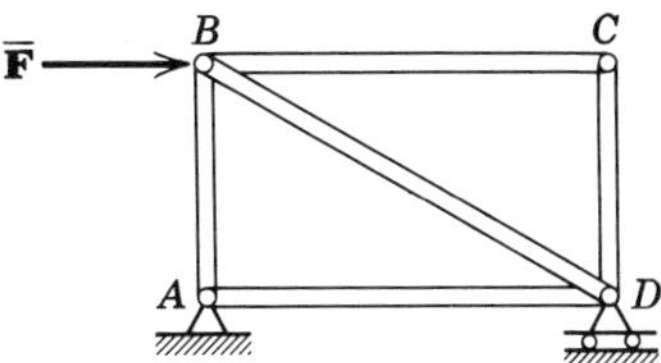

Figure 7.2

that the removal of any member of this truss will allow some part of the truss to collapse. This structure is called a *just-rigid truss.* This truss has another important characteristic. A small change in the length of any member can be made without requiring a change in length of the other members. The other members will adjust their positions to accommodate the change. On the other hand, if we add another member, from *A* to *C*, the situation is quite different. Now any change in length of any member requires a change in length of at least one other member. This system is called *over rigid.* We will limit our consideration to that class of simple trusses called just-rigid trusses.

We define a plane simple truss as a truss that can be constructed in the following way. First begin with a triangle as in Fig. 7.3. The triangle is clearly just-rigid and consists of three joints and three members. Now add additional pairs of members connected by

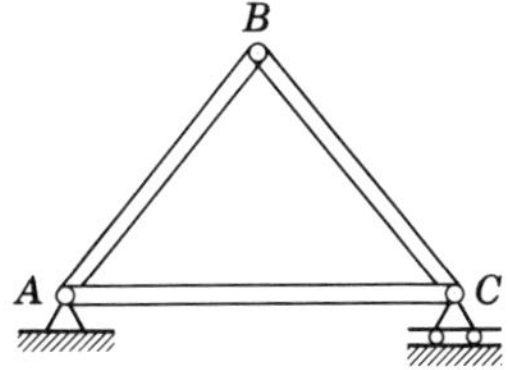

Figure 7.3

one joint. For example, add the two members *BD* and *CD* shown in Fig. 7.4. We can continue to enlarge the truss by adding additional pairs of members connected by single joints.

Let's use the example of Fig. 7.4 to show that

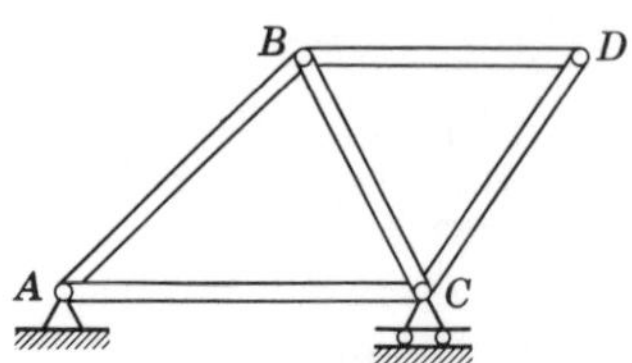

Figure 7.4

there exists a relationship between the number of joints and the number of members in a simple truss. After the initial triangle *ABC*, we must add two additional members for each additional joint. Therefore, the relationship has the form $m = 2j + \text{constant}$, where m is the number of members, j is the number of joints, and the constant we will now determine. For triangle *ABC* alone, $m = 3$ and $j = 3$; therefore, for the equation to hold, for this simplest plane truss, the
−3 constant must equal ____.

The equation is $m = 2j - 3$. Check this result on
5 the truss in Fig. 7.4. Here we have __ members,
4 is __ joints and, therefore, the equation ______ (is/is not) satisfied.

Figure 7.5 shows a free body diagram of the truss of Fig. 7.4 loaded with the known force $\bar{\mathbf{F}}$. How many

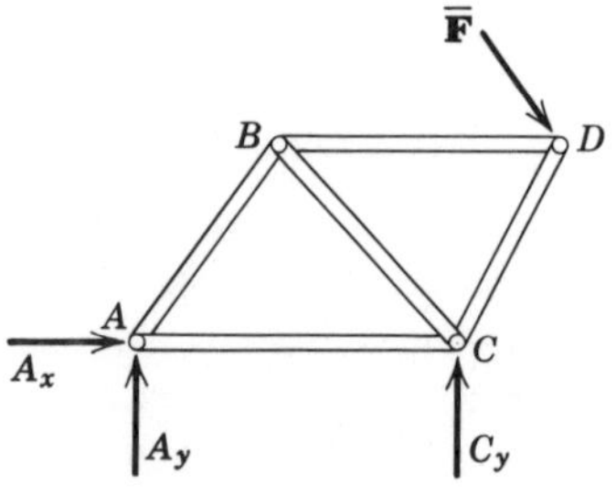

Figure 7.5

unknown quantities are present? We have two reaction components A_x and A_y, the reaction component C_y and the forces in each of the five members. For a truss, each member is a two-force member. The forces in the members must be directed along the line joining their ends. Thus the internal force in each member represents ________ scalar unknown(s). one
(number)

We have a total of 3 unknown reaction components, and __ unknown scalar forces in the members. 5
The total number of unknowns is __. 8

Can we obtain sufficient equations to solve this problem? Consider the equilibrium of the free bodies of each of the joints *A*, *B*, *C*, and *D*. Figure 7.6(*a*) shows

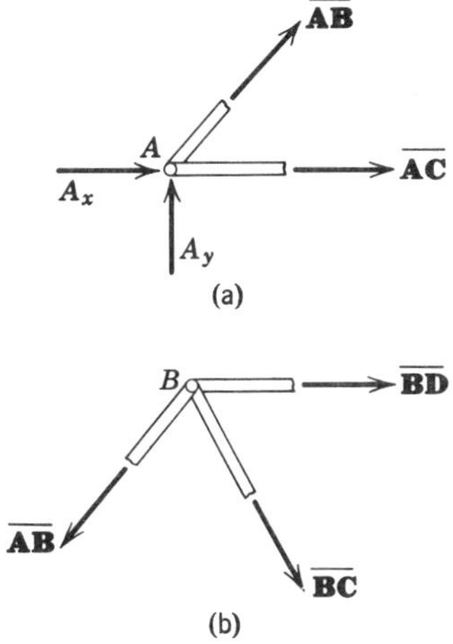

Figure 7.6

the free body of joint *A*. The unknown force *AB* acts along the line through *A* and __. The unknown force *B*
AC acts through *A* and __. *C*

Since we do not know whether these internal forces are tensile or compressive, we assume and draw them as tensile. If our solution is negative, then the force in that member is actually ________. compressive

We can also draw free body diagrams of the other joints such as *B* in Fig. 7.6(*b*). For each joint, the loads form a ______ force system. Therefore, we can write concurrent
(type)

2 ______ independent equations of equilibrium for (number) each joint.

eight Thus we can write a total of ___ independent equations of equilibrium by considering all of the joints.

By consideration of the equilibrium of each joint, we arrive at the exact number of equations needed to solve for the unknowns in a simple truss. We shall now analyze a simple truss. Figure 7.7 shows a truss sub-

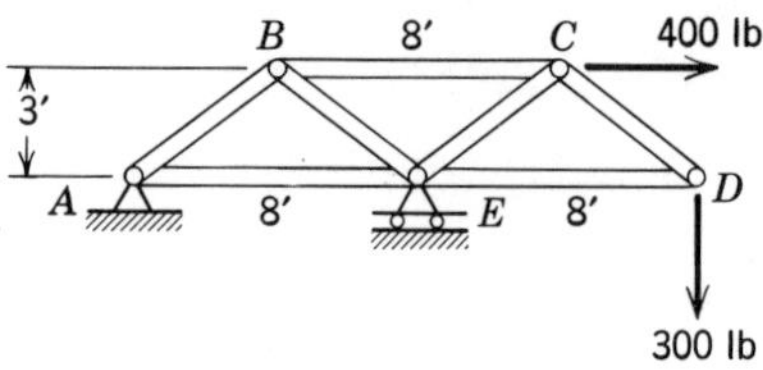

Figure 7.7

jected to external loads. We seek the reactions and the forces in all members. What is our first step?

To draw a free body diagram ______________________________

A suitable diagram is shown in Fig. 7.8. We can

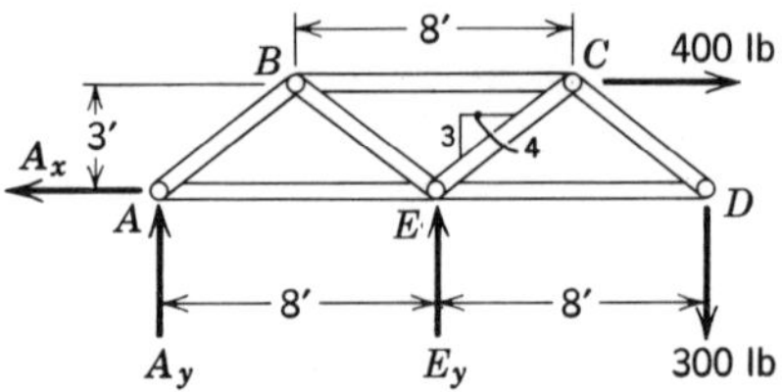

Figure 7.8

begin either by writing equilibrium equations for the entire truss (to solve for the reactions) or by writing equilibrium equations for each individual joint. We will consider the equilibrium of joints individually. It is convenient to begin at a joint with only two unknowns. Thus we avoid larger numbers of simultaneous equations and need only solve two equilibrium

two equations for the ___ unknowns. Our obvious choice

D is joint ___.

Draw a free body diagram of joint *D* in the space provided and check your diagram with Fig. 7.9. The

vertical equilibrium of joint *D* requires that (3/5) *CD*
– _____ = 0. 300 lb

Horizontal equilibrium requires that *DE* + (__). 4/5
CD = 0.

Solving these equations yields *CD* = ___ lb and 500
DE = ____ lb. –400

With the forces *CD* and *DE* known, we seek another joint with only two unknowns. Since *CD* is now
known, a new joint that has only two unknowns is __. C

Consider the free body diagram of joint *C* (Fig. 7.10). On scratch paper, write and solve the appropriate equilibrium equations for this joint to determine *BC* = _____ and *CE* = _____. 1200 lb. –500 lb

Now, since *BC* and *CE* are known, joint __ has B
only two unknowns.

Draw a free body diagram of joint *B* and write and solve the equations of equilibrium to find that
the force in *AB* equals ___ lb and the force in *BE* 750
equals ____ lb. –750

You can now write two equilibrium equations for joint *E* and solve for the unknowns *AE* and E_y. Then you can write two equations expressing the equilibrium of joint *A* to solve for the unknowns A_x and A_y.
Do this using scratch paper to find that *AE* = _____, –200 lb
E_y = _____, A_x = _____, A_y = _____. 750 lb 400 lb –450 lb

Our results can be checked by considering the equilibrium of the entire truss. Summing moments about point *B*, would the reactions, A_x, A_y, and E_y
provide overall equilibrium? _____. Yes
Yes/No

In this problem we were able to work through the truss from joint to joint, solving for two unknowns at each step. Thus we proceeded by solving simple sets of equations. To save time in problem solving,

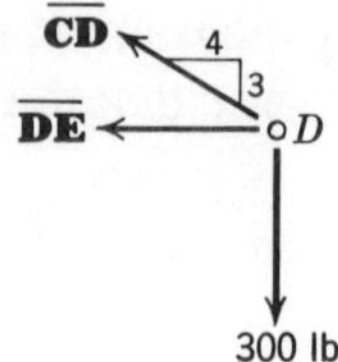

Figure 7.9

one should make use of those situations where one or two equations in one or two unknowns can be written. These can be found by considering the in-

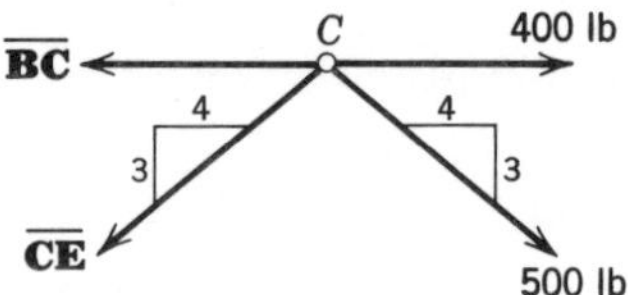

Figure 7.10

dividual joints or the equilibrium of the entire truss. This approach, involving equilibrium of individual joints, is called the *method of joints.*

We next examine several special loading conditions for joints. The joint shown in Fig. 7.11 has no

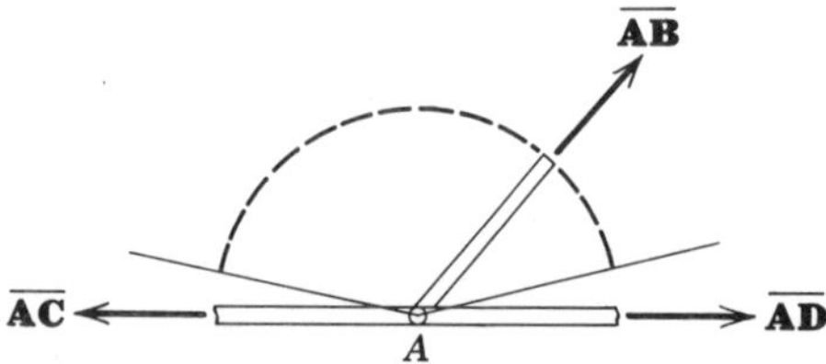

Figure 7.11

external force. Since *all members except one* lie along the same line, the only force that can have a component perpendicular to this line is *AB*. The requirement for force equilibrium perpendicular to

zero this line leads to *AB* = ____.

We conclude that whenever all members at a

joint, except one, lie along some straight line, then the force in that member must equal ____. zero

In the three-dimensional case when all members of an unloaded joint except one lie in a plane, then equilibrium perpendicular to that plane requires that this one member have ____ force in it. zero

Why would a zero force member be present in a truss? Such members are added to prevent buckling. Thus far we have treated only plane trusses. We now consider three-dimensional trusses which are commonly called space trusses. Similar assumptions are made regarding space trusses as plane trusses. For the space truss, however, the tetrahedron rather than the triangle is the basic element. Let us consider now the construction of a simple space truss. To construct a simple or just-rigid space truss, begin with a tetrahedron as shown in Fig. 7.12. Then add sets of three

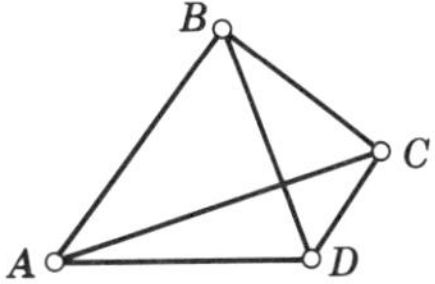

Figure 7.12

members pinned at a single additional joint, as shown in Fig. 7.13. These sets must be connected to the original structure at joints. Thus we enlarge a space truss by the addition of a series of ________. tetrahedrons

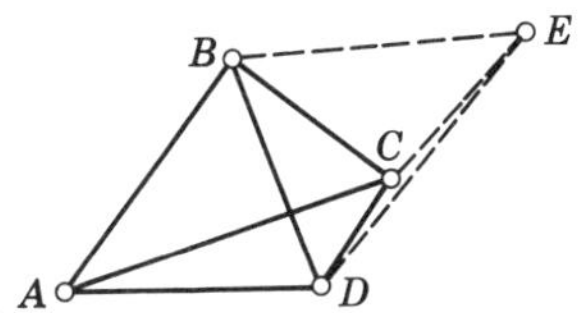

Figure 7.13

There is a relation between the number of members and the number of joints in a simple space truss. Since we add three members for each joint added, the form of the equation must be $m = 3j +$ constant.

The constant is determined to make this relation true for any simple space truss. For the original tetrahedron in Fig. 7.12, with six members and four joints,
−6 the constant must equal ___.
$3j - 6$ The relation for space trusses is $m =$ ___.

The principal difference between space trusses and plane trusses is that with space trusses we are concerned with the equilibrium of three-dimensional systems and with the equilibrium of joints which represent three-dimensional concurrent force systems. For the *entire* space truss we can write six scalar equations of equilibrium. For each joint taken
three as a free body, we can write ___ meaningful equilibrium equations.

There is no essential difference between space trusses and plane trusses. We shall not treat space trusses further here.

We now examine another technique for the solution of plane trusses. Suppose that for the truss shown in Fig. 7.14 you are asked to determine the

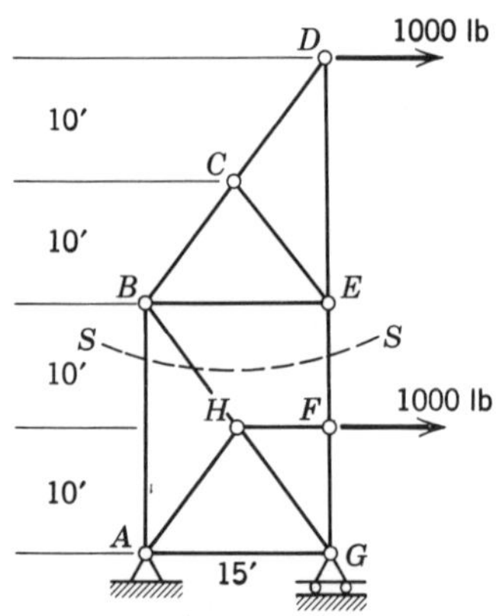

Figure 7.14

forces in members *EF* and *AB*. We would like to avoid solving for the loads in all the other members, as is necessary with the method of joints. Our approach will be to consider the equilibrium of a portion of the truss (Fig. 7.15). Here we draw a free body diagram of a portion of the truss formed by cutting *a section S-S* through the truss. For this free body we

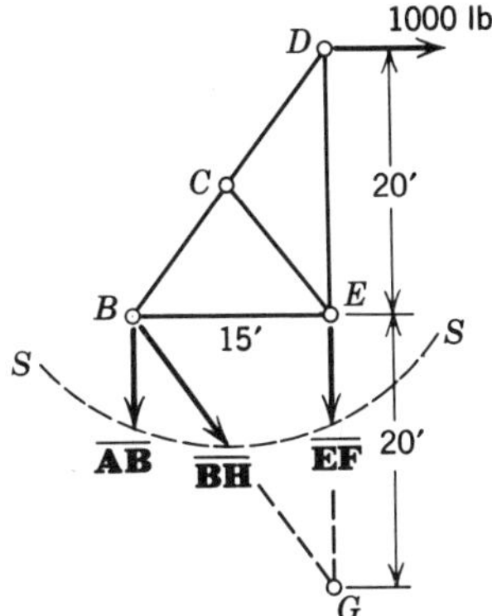

Figure 7.15

do not have a concurrent force system. Therefore, we can write both force and ______ equations of equilibrium. moment

How can we solve for the unknown forces *AB* and *EF*? Is it possible to write a single equation for the free body shown with only the unknown *EF* present? This can be accomplished by writing a moment equation about point __. *B*

This moment equation reads $M_B = -1000 \cdot 20 - 15 \cdot (\_\_) = 0$. *EF*

Solve this equation to find that *EF* = ______ lb. -1333

Notice that by cutting a section through the truss and taking equilibrium of a part of the truss on one side of the section, we are able to solve directly for the force in an internal member. This method is very convenient when the forces in only a few members of a truss are sought.

Next, solve for the force in member *AB*. We write a moment equation about a point such that the only unknown present is *AB*. An equation for the sum of the moments about point __ accomplishes this. *G*

A moment equation can be written about any point in space, not just points on one side of section *S-S*. Now write a moment equation about point *G* and solve to find that *AB* = ____ lb. 2667

We have now solved the problem and found the forces in members *AB* and *EF*. We could, of course, continue to solve for additional members. For example, we could solve for the force in member *BH* by considering horizontal or vertical force equilibrium of this section of the truss.

Often it is desirable to combine this second method, called the *method of sections*, and the method of joints. For example, we might find the forces in members *AB* and *BH* as just indicated and then consider equilibrium of joint *B* to find the forces in members ___ and ___.

BC *BE*

There are several other points to note regarding the method of sections. Do not limit yourself to any one technique in trying to solve problems. Remember that the most efficient solution often involves a combination of equations determined by consideration of the equilibrium of free body diagrams of the entire ____, of various individual _____, and of portions of the truss called ______.

truss joints sections

In addition, there are some very effective graphical techniques of analysis that can be employed for plane trusses. A technique called the *Maxwell Diagram* can be used to advantage when drafting equipment is available. Graphical techniques are covered in a number of standard texts and are not discussed further here.

We now examine another class of simple structures, *frames*. Frames are assemblies of pin-connected members. The principal distinction between frames and trusses is that frames include members which have more than two forces acting on them. Frame analysis involves straightforward application of equilibrium principles. Consider the plane frame in Fig. 7.16. There are more than two forces acting on all members except __.

DE

We seek the forces, including reactions, acting on each member. We begin by drawing a free body diagram of the entire frame, Fig. 7.17. Note that there are ________ unknown scalar reaction compo-
(number)
nents. Can we expect to solve for all of them, using equations for this overall free body diagram alone? ______.
Yes/No

four

No

We can solve for *some* of them, however. For example, by taking moments about point *G*, we can find the reaction component __. Taking moments

A_y

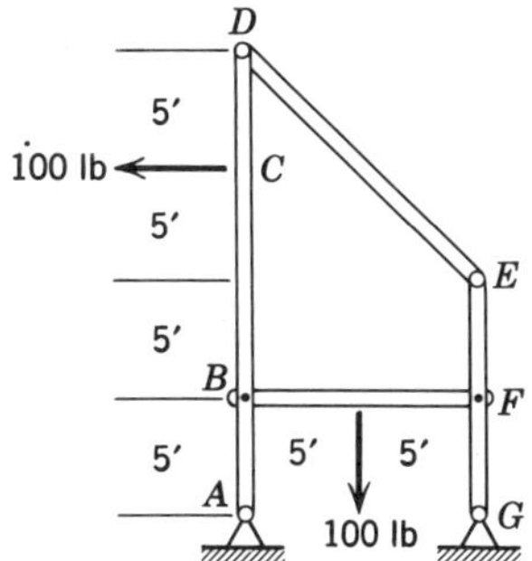

Figure 7.16

about point A, we can find the unknown component ___. G_y

Perform these operations and find that $A_y =$ ____ 200 lb
and that $G_y =$ ____. −100 lb

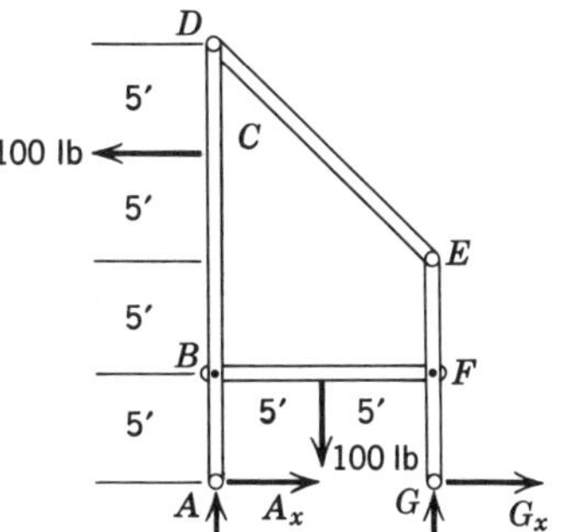

Figure 7.17

To determine the other reaction components and the forces acting on the various members, it is necessary to consider the members individually. Fig. 7.18 shows free body diagrams of the individual members. Note that the forces acting at each joint *must* be shown in a manner consistent with Newton's third law. For example, if the force components acting on member *AD* at *B* are taken as positive upward and to the right, then the corresponding forces on member *BF* at *B* must be shown with components downward and to the left. Similarly, if the force components

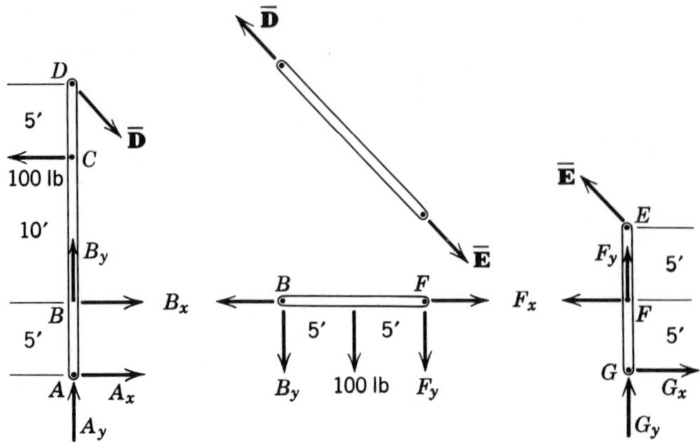

Figure 7.18

acting on BF at F are positive downward and to the right, the force components on EFG at F must be
upward left directed ______________ and to the ________.
upward/downward left/right

Since DE is a two-force member, the forces acting at each end act along the line DE. Let us now determine as many of the unknown force components of the system as possible without solving any simultaneous equations. First, we can determine B_y and F_y by writing moment equations for member BF about
−50 lb points B and F. Do this and find that B_y = _____ and
−50 lb F_y = _____.

We have now determined the values of the unknown force components A_y, G_y, B_y, and F_y. Circle these known quantities in Fig. 7.18. There remain
six to be found A_x, B_x, D, E, F_x, and G_x, a total of __ unknowns.

There is no other equation involving only one unknown. We seek an efficient way of finding these unknown forces. Consider the free body of member
$\bar{\mathbf{E}}$ DE. The force $\bar{\mathbf{D}}$ equals the force __.

Also, from force equilibrium of member BF, the
F_x force component B_x equals __.

Another useful relationship is obtained from horizontal force equilibrium of the overall structure as
100 lb shown in Fig. 7.17. Here we find that $A_x + G_x$ = _____.

To solve for the remaining unknown force components, we must write a set of simultaneous equations. One approach is to take moments about point D for member *AD* and about point *E* for member *EG*. These two equations, combined with the two equations relating to B_x, F_x, A_x, and G_x, can be solved for these components. Therefore, M_D (for *AD*) = ______________ ______ and M_E (for *EG*) = ____________.

We have four simultaneous equations relating the unknowns A_x, B_x, F_x, and G_x. Solve these equations to find that A_x = _____, B_x = ______, F_x = ______, and G_x = ______.

Only the unknowns *D* and *E* remain. Summing horizontal forces for member *EG* yields *E* = ______ _____. Therefore, *D* = _____.

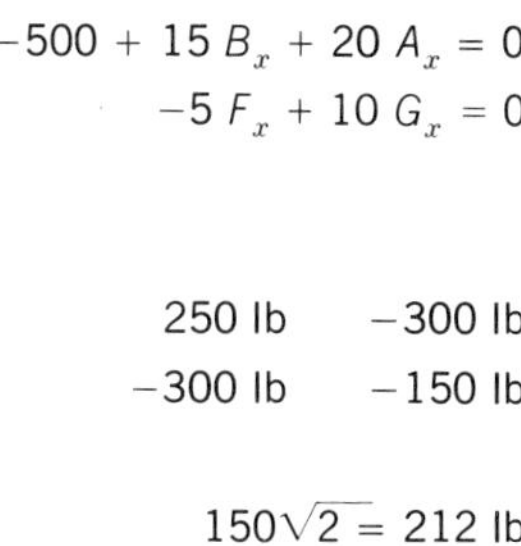

$-500 + 15\,B_x + 20\,A_x = 0$

$-5\,F_x + 10\,G_x = 0$

250 lb   −300 lb

−300 lb   −150 lb

$150\sqrt{2} = 212$ lb

212 lb

We now have all external reactions and internal forces at the joints of this truss. Let us now verify our results. Check the sum of the moments about point *A* for member *AD* and demonstrate that equilibrium is satisfied.

There exist statically indeterminate frames as well as frames which are not rigid; however, we limit our study to just-rigid frames which can be analyzed using the techniques of statics. Since there are no simple tests available to determine if a frame is just-rigid, one must rely on intuition. For further information on the analysis of trusses and frames, see any standard text on structural analysis.

7.2 Beams

Beams are useful primarily because their resistance to bending allows them to support lateral loads over a span. Our study of beams will require some knowledge of the internal forces acting in structural members. The beam in Fig. 7.19, is built in (rigidly

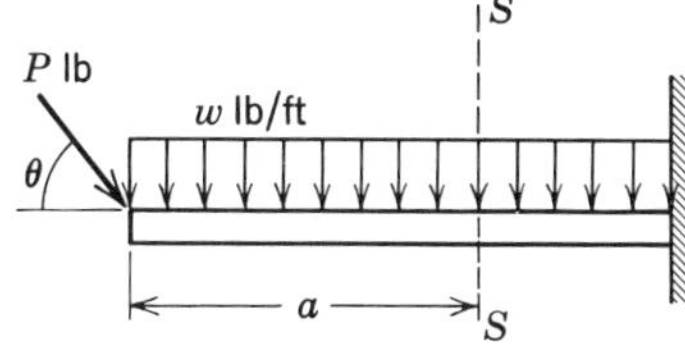

Figure 7.19

fixed) at the right end. It supports a distributed load w lb/ft and a force P at angle θ on its free end.

How can we determine the internal forces and moments that are transmitted across the beam section S-S located the distance a from the free end? First, we draw a free body diagram of the beam to the left of the section S-S. The forces acting on the area elements of the cross section S-S will vary, some being tensile and others compressive. Determination of the distribution of these forces is beyond the scope of this course. However, the resultant of the distribution must provide equilibrium for the part of the beam considered. The resultant force and resultant moment of the force distribution on the section S-S must be such that the sum of the forces and moments for the entire free body will equal ___. (zero)

Call the resultant moment, on the section S-S, M, and the resultant force $\bar{\mathbf{F}}$, then split the force into two components, one vertical and one horizontal. In Fig. 7.20 the free body diagram of the beam section

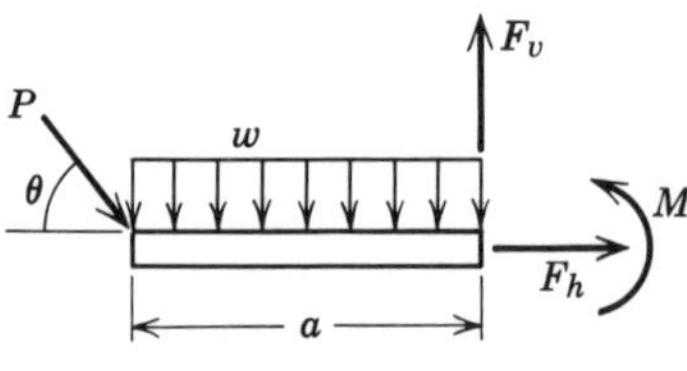

Figure 7.20

with the resultant force and moment on S-S is drawn as specified. The equations of equilibrium hold for this body. Summing horizontal forces leads to the equation $P \cos \theta + \_\_ = 0$. (F_h)

Vertical equilibrium requires that $F_v - wa - \_\_\_ = 0$. ($P \sin \theta$)

To determine M, take moments about the section S-S and obtain the equation: $M + aP \sin \theta + w(\_) = 0$. ($\frac{a^2}{2}$)

Note that the axial force F_h, transverse (*or shear*) force F_v, and bending moment M, at *any* cross section can be determined by equilibrium considerations and are functions of the distance a along the beam. Since the stresses and deflections of laterally loaded beams depend upon their shear and bending moment distributions, we shall examine these concepts further. Since the axial forces in beams are usually unimportant, we shall omit them.

First we must establish a sign convention for internal shear and moment. Figure 7.21 shows a loaded beam that has been cut at section S (the two sections are separated for clarity). We adopt the directions shown as positive directions for moment M and shear V. These conventions define shear force as *positive* if it is upward on the *left* end of an element and a *positive* moment as one that produces *compression* in the top of the beam. Due to Newton's third law, if a shear (shear force) upward on the left end of an element is positive, then a shear ________________ downward

upward/downward

on the corresponding right end must be positive (as shown in Fig. 7.21).

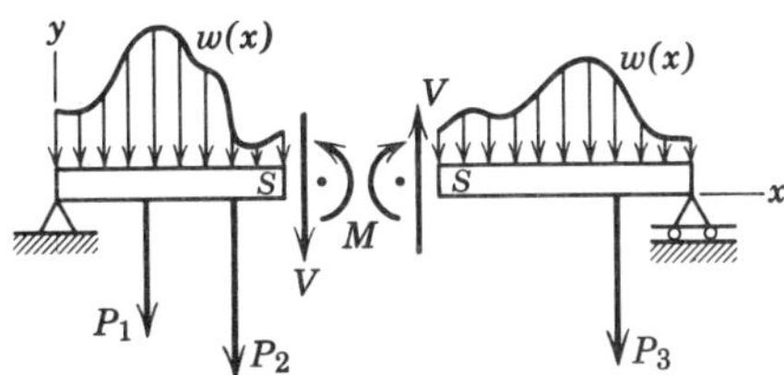

Figure 7.21

We also adopt the convention that a distributed load, w, is positive *downward* (in the negative y direction). We have established sign conventions for shear, moment, and lateral load. Next, we shall determine some relations between these quantities, for the case in which all loads lie in a single plane. We examine a small segment, of differential length, cut from a beam (Fig. 7.22). Since the shear and moment may vary along the beam, we have shown the moment and shear at the cross-section $x = a$ as M and V and the moment and shear at the cross-section $x = a + \Delta x$, as $M + \Delta M$ and $V +$ ____. ΔV

The vertical force equilibrium of this element is expressed by the relation $V - w\Delta x - (V + \Delta V) = 0$.

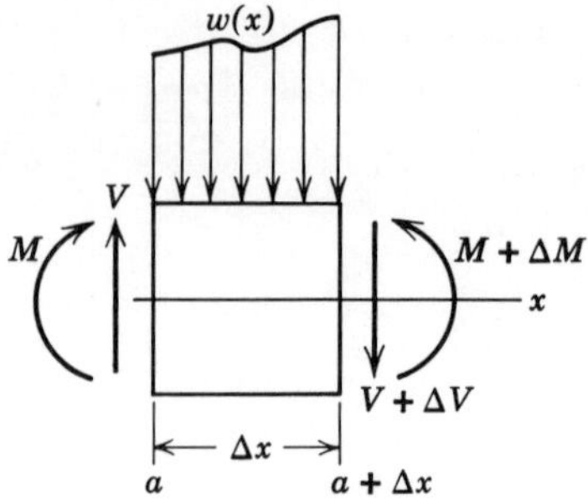

Figure 7.22

Cancel the appropriate terms, divide through by Δx, and obtain the relation ______________.

$-w - \Delta V/\Delta x = 0$ or $-w = \Delta V/\Delta x$

If we take the limit of this expression as Δx approaches zero, we obtain ______.

$-w = dV/dx$

This relation is an elementary differential equation relating V, x, and w.

To obtain a relation involving the moment, we consider moment equilibrium of the element of Fig. 7.22 about any point on the section at $a + \Delta x$. This equation (retaining only first-order terms) reads $-M - V\Delta x + w\,(\Delta x)^2/2 + (\_\_\_\_\_) = 0$.

$M + \Delta M$

Cancel appropriate terms, divide by Δx, and take the limit as Δx approaches zero! We find that V = ____.

dM/dx

We have obtained the relations, $dV/dx = -w$ and $dM/dx = V$. We interpret these relations as follows: the slope (dV/dx) of the shear function must equal the negative of the distributed load intensity (w) and the slope (dM/dx) of the moment function equals the value of the shear (V) at each point.

Concentrated loads have not been included in these derivations. Therefore, these relations are valid only on uniformly loaded (or unloaded) portions of a beam between concentrated loads. The behavior of the shear at a concentrated load can be determined from the relation $dV/dx = -w$. Consider the concentrated load as the limit of a distributed load as its intensity increases. Thus at a concentrated load dV/dx becomes infinite, that is, there is a jump in the shear.

Similarly, we can show that a concentrated mo-

ment produces a jump in the ____________ function
shear/moment
and a concentrated force produces a change in the slope of the moment.

moment

Let us use these results to construct diagrams of shear and moment distributions. For example, consider the cantilever beam of Fig. 7.23. It is subjected

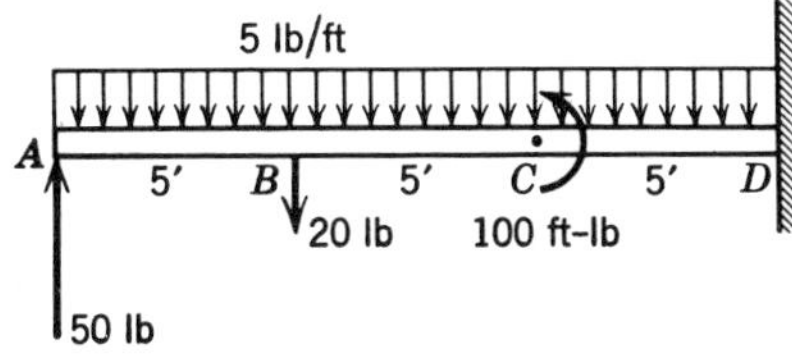

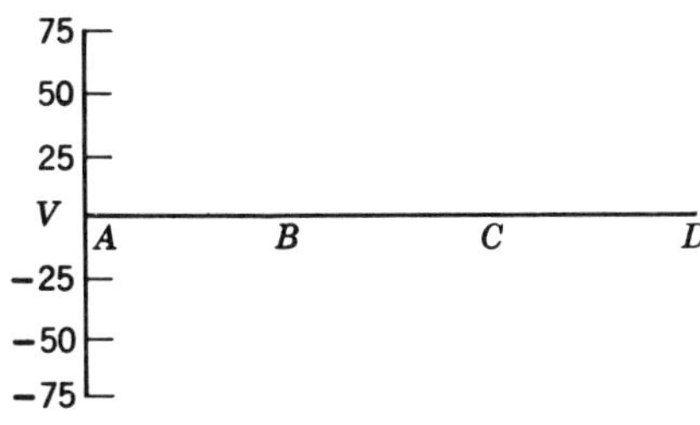

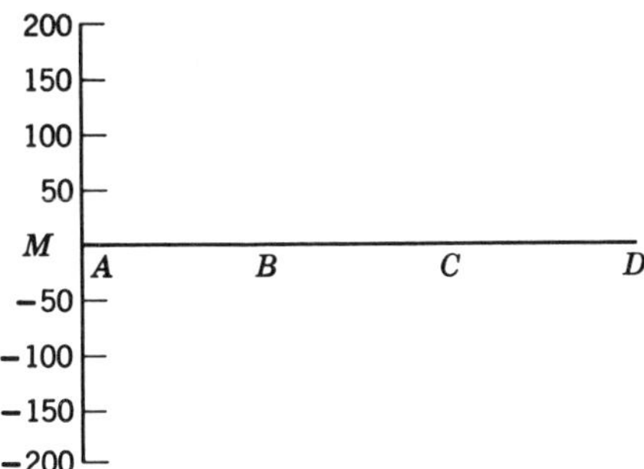

Figure 7.23

to a uniformly distributed load, several concentrated loads, and a concentrated moment. We will draw the

shear distribution beginning at the left end of the beam. We have a 50-pound force acting upward on the left end of the beam. Thus the shear at point A is 50 pounds. Since this shear force is upward on the left end, it is a ______________ shear.
positive

negative/positive

Plot this point on Fig. 7.23.

Between the concentrated loads at A and B we use the differential relation $dV/dx = -w$. The load function $w(x)$, which is a constant, equals five pounds per foot. Therefore, the slope of the shear curve, dV/dx, will be constant and have the value ______. Of course, a curve with constant slope is a ______ line.
−5 lb/ft
straight

Draw on the diagram a line starting from $V = 50$ with a slope of -5 pounds per foot. If done correctly, this line will lead to a value $V =$ __ at point B. Plot this point on Fig. 7.23.
25

A concentrated force of 20 pounds acts at B. At a concentrated force there is a ____ in the shear diagram.
jump

Consider the beam element in Fig. 7.24. By our

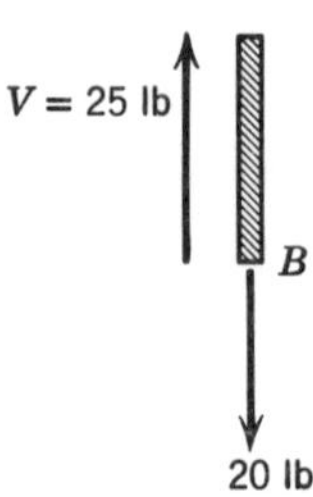

Figure 7.24

sign convention, the shear on the left of the element is a positive 25 lb (upward on the left). For vertical force equilibrium there must be a shear force ______________ on the right of magnitude 5 lb.
downward

upward/downward

Draw this 5-lb shear force onto Fig. 7.24. This shear is ______________ (downward on the right).
positive

positive/negative

Thus, in crossing the element, the shear has changed by 20 lb, from 25 lb to 5 lb. Since this is a differential element, this change is drawn as a jump in the shear function at point B. Plot this value at point B on the diagram Fig. 7.23. The shear is continuous for the entire section BCD, since there are no concentrated ____ in this region. The concentrated moment at C contributes a jump in ______ but does not produce a jump in shear.

forces

moment

For section BCD, the slope of the shear function, $dV/dx = -w = -5$ lb/ft. Draw this line on Fig. 7.23, continuing from the value $V = 5$ lb at B. Since the slope is -5 lb/ft and the distance from B to D is 10 feet, the intercept or shear value at D is $V =$ ___ lb.

−45

We can check the value of the shear at D by considering equilibrium of the entire beam. Do this to prove that the reaction force at D equals 45 lb.

We will now draw a diagram of the moment distribution in the beam. At A there is no external moment. Therefore, the moment diagram has a value at A of ___ ft-lb.

zero

We shall use the shear diagram in our construction of the moment diagram. A correct shear diagram is shown in Fig. 7.25. Examine the relation $dM/dx = V$.

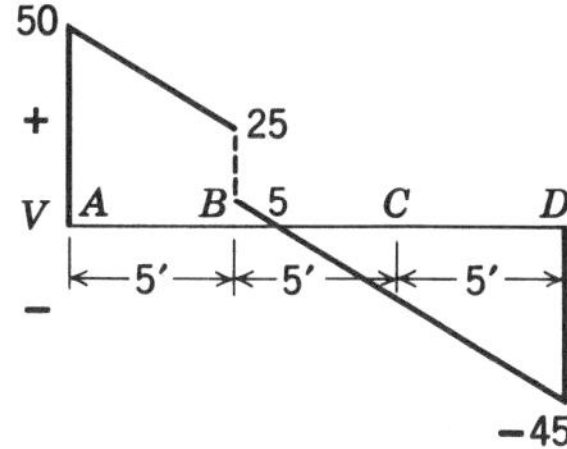

Figure 7.25

This relation states that the slope of the moment diagram equals the value of the shear at each point. From the shear, we see that the *slope* of the moment diagram is positive at A because the shear is positive. At B there is a jump in shear and thus a sudden change in the slope of the moment diagram. Furthermore, the moment attains a maximum where the shear is zero. The concentrated moment at C produces a ____ in the moment diagram.

jump

Write $dM = V\,dx$. For section AB of the beam we know V as a continuous function of x. Thus we can integrate the relation to find $M_B - M_A = \int_{x=0}^{x=5} V(x)\,dx$. This equation states that the *change in moment* between two points *equals* the *area* under the shear curve between those points. From Fig. 7.25 we see that the shear in region AB is $V(x) = 50 - 5x$ lb. Substitute for $V(x)$ in the integral. Integrate to show that $M_B - M_A$ = ______.

$(50x - 5x^2/2)\,\big|_0^5$

Notice that the moment curve between A and B is parabolic. We now know the following information about the moment curve to be drawn between A and B. The moment is zero and has a positive slope at A. The slope decreases from A to B, since the shear decreases. Furthermore, the shape of the curve is parabolic.

The value of moment at B can be obtained from the expression in the previous frame. Insert the limits to find M_B = ____ ft-lb.

187.5

With this information, *sketch* the moment curve between points A and B in Fig. 7.23.

Next, consider the section from B to C. Write an expression for the shear in this section *using a new coordinate system, measuring x from B*, as $V(x) = 5 -$ __x lb.

5

Integrating, we find that the change in moment $(M_C - M_B)$ between B and C is given by the expression $(5x -$ ___$)$ evaluated between $x = 0$ and $x = 5$.

$\frac{5x^2}{2}$

The moment M_B is 187.5 ft-lb. Evaluate the previous expression. Thus the moment $M_C = 187.5 -$ ___ = ___ ft-lb.

37.5

150

We have the following information for section BC. The moment at B is 187.5 ft-lb, the slope of the moment curve at B is positive, and at C the moment is 150 ft-lb. At C the slope of the moment curve is ______ (from the shear value).
positive/negative

negative

The moment reaches a ______ in section BC at the point where the slope of the moment curve and the value of the ____ are zero.
maximum/minimum

maximum

shear

The location of this maximum is found by determining the point of zero shear. Consider Fig. 7.25

and use similar triangles; how far to the right of *B* would the maximum occur? ______. One foot

This value, inserted into the moment equation for section *BC*, $M(x) = 187.5 + 5x - (5x^2/2)$, yields M_{max} = ______. 190 ft-lb

The concentrated moment at *C* produces a jump in the moment diagram. The differential element at *C*, Fig. 7.26, has a positive 150 ft-lb moment on its left

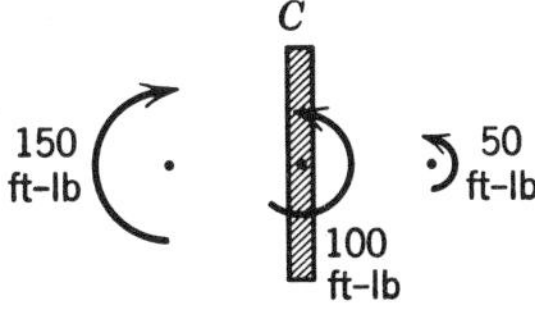

Figure 7.26

side. For moment equilibrium there must be a 50 ft-lb moment on the right, which is ______ (positive/negative) as shown. positive

We thus have a moment value of ______ just to the right of *C*. Plot this value in Fig. 7.23. +50 ft-lb

To complete our moment diagram, consider the shear function for the section *CD*. *Using a coordinate system that measures x from C*, an expression for the shear in *CD* can be written as $V(x)$ = ______ lb. $-20 - 5x$

Integrate this expression to show that the change in moment between points *C* and *D* equals ______ ft-lb. -162.5

Again the moment curve is parabolic. Sketch it in the space of Fig. 7.23. Check your result with Fig. 7.27.

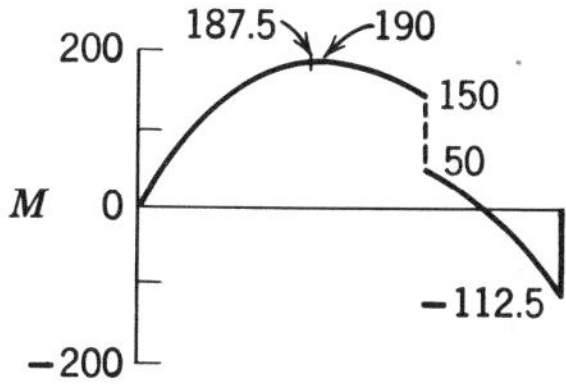

Figure 7.27

A fact useful in the construction of moment diagrams is that the change in moment between two points equals the area under that section of the shear diagram. To demonstrate this result consider section *CD* of the beam. The shear at *C* is −20 lb and at *D* is −45 lb. The distance involved is five feet. The area between the zero shear axis and the curve equals

−162.5 below

_____ ft-lb and is negative, since it is __________
below/above
the *x* axis.

Check the moment at *D* by applying moment equilibrium to the entire beam.

Shear and moment diagrams can be constructed by considering the *equilibrium* of sections of the beam. It is generally simpler, however, to use the differential relations as illustrated step by step. For beams supported at both ends it is necessary to solve for the reactions before constructing shear and moment diagrams.

7.3 FLEXIBLE CABLES

Cables and chains are characterized by their relatively small resistance to bending. As an engineering approximation, we neglect bending resistance and consider only the tensile forces involved. We will limit our investigation to those problems where all loads and the cable lie in a plane. Three-dimensional problems, though complex, involve the same principles.

Consider a flexible cable that is loaded at discrete points and for which *the weight of the cable is negligible* compared to the external loads. Since the cable

straight

cannot resist bending, it is ____________ between
curved/straight
load points and supports. There can be no bending moment in a flexible cable. Thus, there can only exist forces at the ends of each straight section.

As our example, we will determine the tension in each section of the point-loaded cable of Fig. 7.28. Here the support points *A*, *B*, and the lines of action of the external concentrated loads are given. We also

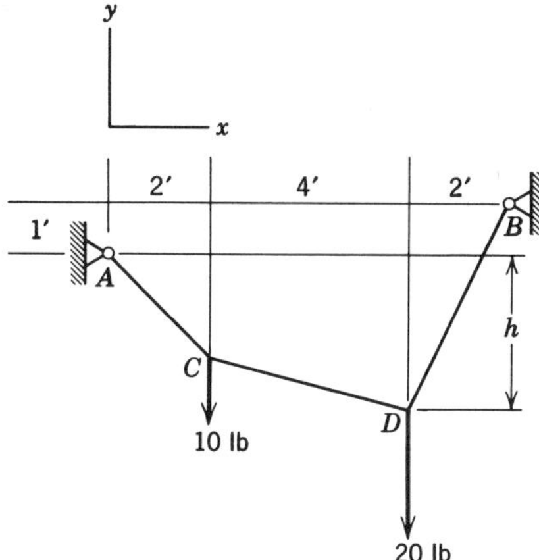

Figure 7.28

need some information regarding the length of the cable. The additional information can be knowledge of the total length of the cable or the vertical location of one of the points of application of a concentrated load. Problems that are formulated in terms of the total length of the cable are algebraically complex. Often, one is concerned only with the sag at some point of the cable (that is, the distance to the cable from some horizontal line). We will consider only cables for which the vertical location of one of the concentrated loads is given. Our additional information for this example is the height, $h = 3$ ft, of the left support above the point of application of the 20-lb load. Draw a free body of the entire cable in the space provided. Now compare with Fig. 7.29.

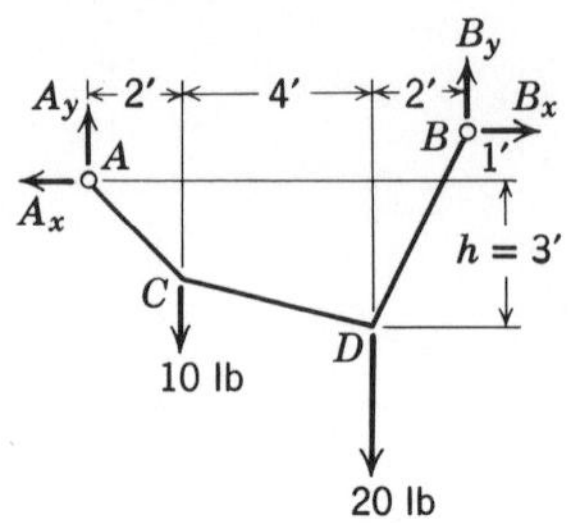

Figure 7.29

If we split the reactions at A and B into two com-
four ponents, we see that there are a total of ______ (number) unknown scalar reaction components in the problem. With only three equations of equilibrium, can we
No solve for these components? ______. Yes/No

Another relation is needed. Since h is given for point D (adjacent to a support) and the cable section DB is a straight two-force member, the ratio of
$B_y/B_x = 4/2$ B_y to B_x is (______).

When the sag is given for an interior point (not adjacent to a support), an additional relation can be obtained by writing a moment equation for a section of the cable to one side of the point of known sag about that point. This method is applicable to any point of known position. Use this technique to verify the relation that $2B_y - 4B_x = 0$.

With this relation and the three equilibrium equations for the overall free body, solve for the four
9.33 lb reaction components. These are A_x = ______, A_y
11.34 lb 9.33 lb 18.66 lb = ______, B_x = ______, and B_y = ______.

With the components at A and B known, it is easy to calculate the other forces of the system. This may be done in a step-by-step manner, working from either end point or by using moment equations. To begin the step-by-step procedure, consider the free body of point B (Fig. 7.30). Write and solve an appropriate force equilibrium equation to find that T_{BD}
20.8 = ____ lb.

This process is continued by considering equilibrium of the load points of the cable.

Suppose that we are asked to solve only for the tension T_{CD} and the position of cable section CD. We will use moment equations written for that part of the cable to the right of point E (located halfway be-

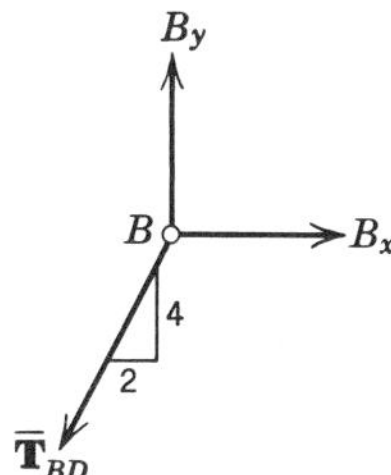

Figure 7.30

tween C and D as shown in Fig. 7.31). Using the values of B_x and B_y found previously, we can find the position of point E, y_E, by summing moments about point E. Do this to find that y_E = ___ ft. 3.72

With y_E known, θ_1 is easily determined. We can now find the magnitude of the force $\overline{\mathbf{T}}_{CD}$ by considering vertical and horizontal force equilibrium for the cable element EDB of Fig. 7.31. Write and solve the appropriate equations and show that T_{CD} = ___ lb. 9.42

Consider the case where all the applied loads are directed *vertically downward*. Examination of a free body of any portion of the cable shows that the horizontal component of the tension must be

________________. constant

variable/constant

The value of this horizontal component of tension is equal in magnitude to B_x or ___. A_x

The tensile force in the cable must be larger than

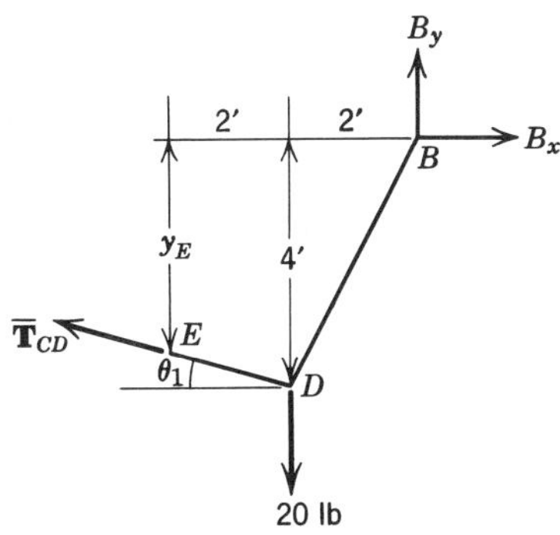

Figure 7.31

B_x (equals A_x) and is maximum where the angle of the cable with the horizontal is ________________. maximum
maximum/minimum

Thus the greatest tension in the cable occurs where the slope of the cable is steepest. This must occur at one of the ______. supports

We will now consider cables that support vertical *distributed* loads. For convenience, we place the origin of our system at the lowest point on the cable, point *C*, Fig. 7.32. The advantage in using this origin

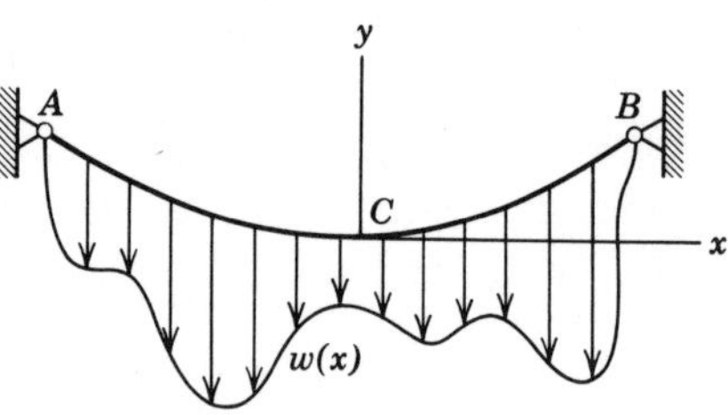

Figure 7.32

is that the tangent to the cable at this point is ______. horizontal

The relation $dM/dx = V$ (developed earlier for beams), is valid for cables. We assume that the moment in a flexible cable is everywhere zero. Thus the relation shows that the shear is zero everywhere and only an axial force may exist. Therefore, the tensile force in a flexible cable is always ______________ to
normal/tangent

the cable itself. tangent

In a vertically loaded cable the *horizontal component* of the tension at any point is a constant, T_0 (equal in magnitude to the horizontal reaction components). At the origin the tension in the cable equals __. T_0

Next consider the free body of a section of the cable to the right of point *C* (Fig. 7.33). $W(x)$ is the resultant vertical load applied to this segment of the cable. Force equilibrium yields several relations of general applicability. For example, the equation of vertical force equilibrium is $T \sin \theta =$ ___. $W(x)$

Similarly horizontal equilibrium yields $T \cos \theta$ = __. T_0

To combine these equations, square both sides

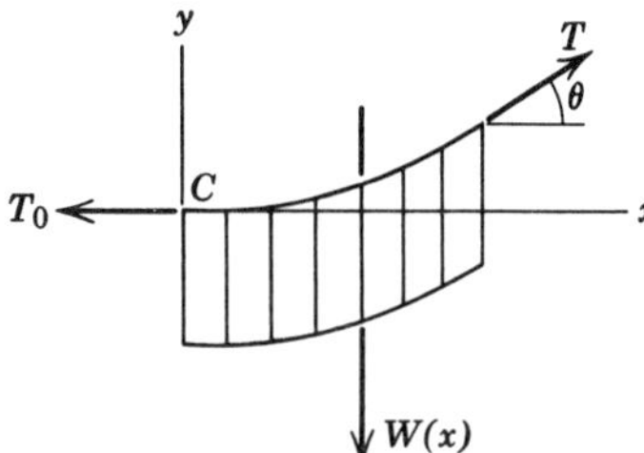

Figure 7.33

and add. A familiar identity simplifies the relation to $T^2 = (\_\_\_)^2 + (\_\_)^2$. $W(x)$ T_o

The division of the first equation by the second shows that $\tan \theta =$ ___. $\frac{W(x)}{T_0}$

Since $\tan \theta$ equals dy/dx (where y is the vertical coordinate of the cable), this relation equals $dy/dx = W(x)/T_0$.

From this elementary differential equation we can find an equation for the shape of the cable curves, provided that $W(x)$ is known and T_0 can be found. An elementary but important example is the cable shown in Fig. 7.34. This cable has a load (w lb/ft) *distributed*

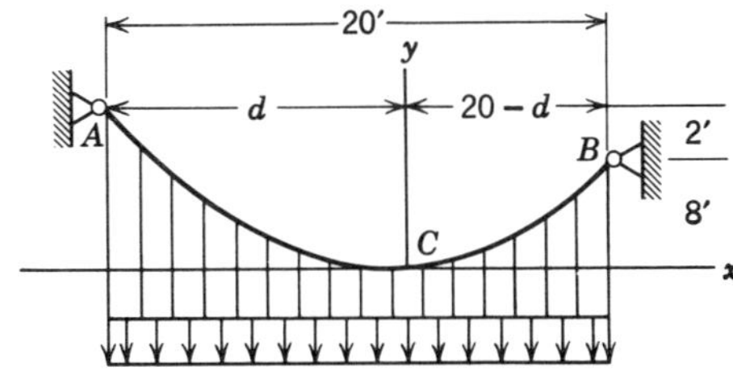

Figure 7.34

uniformly in the x or horizontal direction. We first determine the function $W(x)$ in order to integrate the differential equation and determine y as a function of x. If $W(x)$ represents the total load on a cable segment of length x and w represents the load per unit length in the x direction, then $W(x) =$ __x. w

Rewrite the differential equation as $dy = wx\, dx/T_0$.

$\frac{wx^2}{2T_0}$ + const

Since w and T_0 are constants, we can integrate to find that $y =$ ________.

We have found that *any* cable with the load distributed uniformly in the x direction has a *parabolic* shape. With w given, we can determine T_0 and the integration constant. At x equals zero, y equals zero.

zero

The integration constant is ___.

To determine T_0, we must incorporate information regarding the length or sag of the cable. Since problems formulated in terms of the length of the cable are algebraically difficult and relatively uncommon, we will only consider a formulation that involves the sag of the cable. A description of the treatment of problems formulated in terms of cable length can be found in most standard texts.

Somehow we must use the fact that the lowest point of the cable is a known vertical distance below the supports. The *horizontal* distance from A to C (currently unknown) is labeled d in Fig. 7.34. The horizontal distance from C to B is 20-d.

We now have two unknowns T_0 and d. Using the equation $y = wx^2/2T_0$, we write for point B, $8 = w(20 - d)^2/2T_0$. Similarly, the equation for point A

$wd^2/2T_0$

is $10 =$ _____.

Solve these equations simultaneously and deter-

5.5 w 10.5

mine T_0 and d. $T_0 =$ ___ lb; $d =$ ___ ft.

With this information available, the tension T in the cable at any point can be found. First find the slope at that point from the relation $\frac{dy}{dx} = \frac{W(x)}{T_0} = \tan\theta$. Then use the relation $T\cos\theta = T_0$.

For other vertical load distributions we would find the appropriate function $W(x)$ to use in the relation $dy/dx = W(x)/T_0$. A common example, treated in most standard textbooks, involves a load distributed uniformly *along the length of the cable* rather than *in the horizontal direction*. The curve assumed by the cable for this type of loading is called a catenary.

In this chapter on elementary structural analysis we have simply applied equilibrium considerations to free body diagrams. Many other problems can be readily treated through application of the principles discussed here and in previous chapters.

Summary

A truss is a structure consisting of a collection of pin-connected two-force members in which all forces act at the pins. The triangle is the basic truss element. For simple plane trusses the number of members and joints is related by $m = 2j - 3$ (m—members, j—joints). The method of joints for analyzing trusses involves writing equations of force equilibrium for each separate joint of the truss. The method of sections utilizes force and moment equilibrium equations written for free bodies of portions of trusses containing a number of joints. Often truss problems are solved most efficiently through use of a combination of the two methods.

A frame is a structure consisting of an assembly of pin-connected members, not necessarily two-force members. They are analyzed by the straightforward solution of equilibrium equations written for separate free bodies of the individual members.

Beams are used to support lateral loads over a span. In straight laterally loaded beams, we are primarily concerned with the transverse or shear force and the moment at each cross section. The convention that shear is positive when up on the left end of a beam section and moment is positive when it causes compression of the upper fibers is adopted. Consideration of equilibrium of a beam element leads to the relations

$$\frac{dV}{dx} = -w \qquad \text{and} \qquad \frac{dM}{dx} = V$$

among lateral load intensity, shear, and moment. Shear and moment diagrams are easily drawn using these relations and noting that concentrated forces produce jumps in the shear and concentrated moments in the moment distribution.

Flexible cables and chains have negligible resistance to bending. Cable problems require specification of cable length or the sag at one point, in addition to the location of loads and end points, to be properly posed. Point-loaded cables are straight between loads. Plane cables with distributed loads are described by the relation

$$\frac{dy}{dx} = \frac{W(x)}{T_0}$$

which may be integrated in specific cases utilizing the fact that the cable tangent is horizontal at the lowest point. Cables with loads distributed uniformly in the horizontal direction are parabolic in shape.

Problems

(1) The truss in Fig. P7.1 consists of horizontal and diagonal elements of length L. If only the load in element BC is desired which method is most appropriate? Find the load in BC.

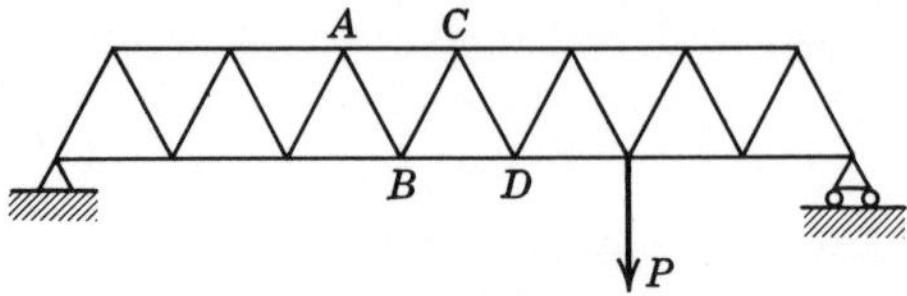

Figure P7.1

(2) Compute the loads in all elements of the truss in Fig. P7.2. Will there be any difference if point A is on rollers and C is rigidly restrained?

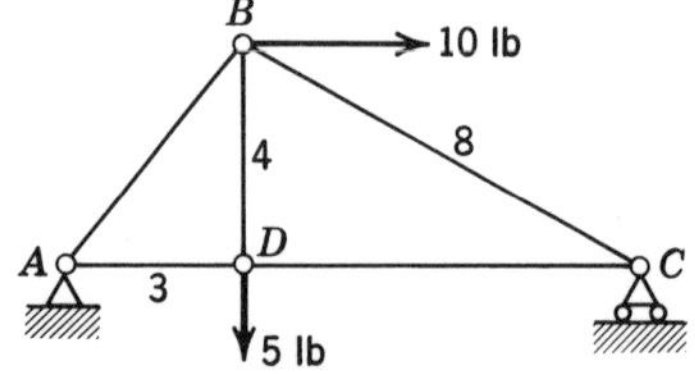

Figure P7.2

(3) A bar is placed between two cantilevered (built-in) beams AC and BD. Find the bending moment and shear force at point A in Fig. P7.3.

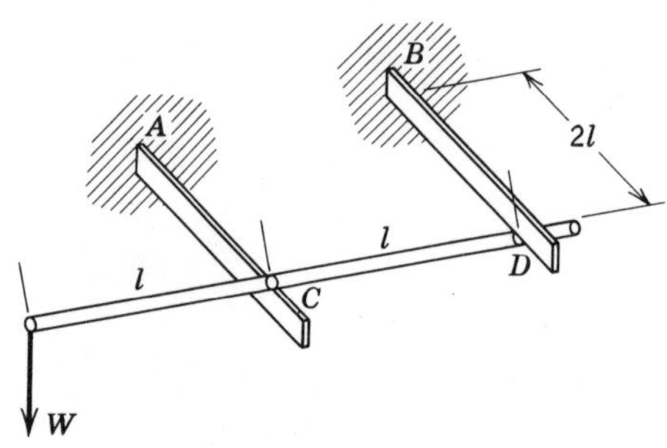

Figure P7.3

(4) In Fig. P7.4 three rods, AB, AC, AD of length 2ℓ are pinned at A and supported by casters at B, C, and D. They are chained together at their midpoints by chains of length $\ell/2$. What is the bending moment at a point just above E caused by the pot of weight w suspended from the apex?

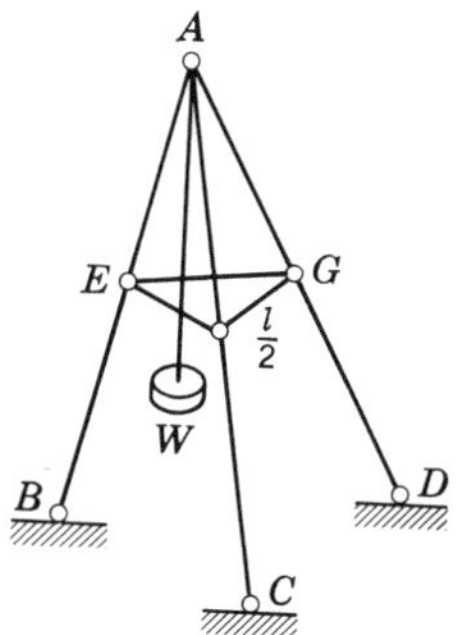

Figure P7.4

(5) A space truss in Fig. P7.5 supports a vertical load P. Determine the forces in members BC and BA.

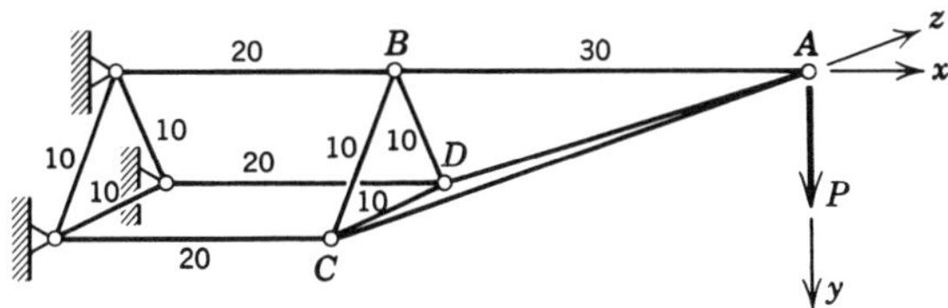

Figure P7.5

(6) Calculate the reactions at the supports of Fig. P7.6.

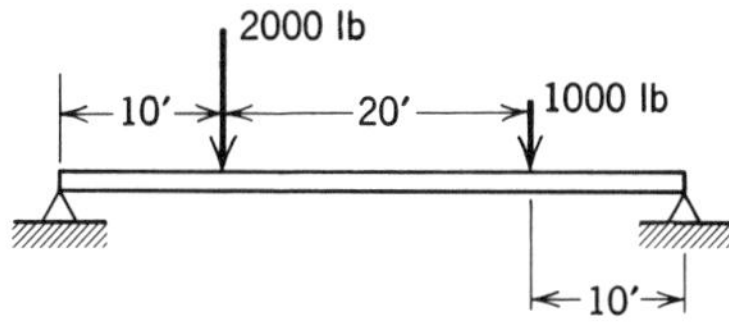

Figure P7.6

(7) Draw the shear and bending moment diagrams for the beam in Fig. P7.6.

(8) Calculate the reactions at the support in Fig. P7.7 and draw the shear and bending moment diagrams for this beam.

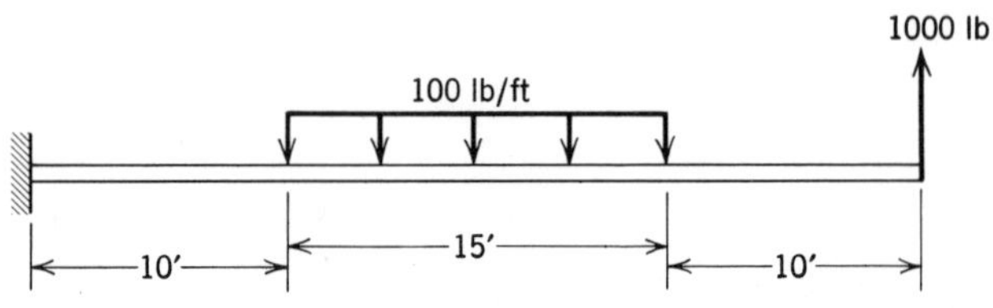

Figure P7.7

(9) Sketch the shear and bending moment diagrams for the beam in Fig. P7.8.

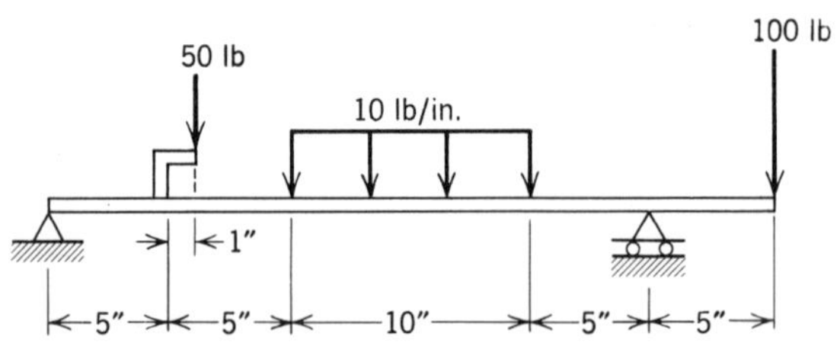

Figure P7.8

(10) If $a = 8$ ft and $b = 10$ ft, determine the components of the reaction at E for the cable and loading shown in Fig. P7.9.

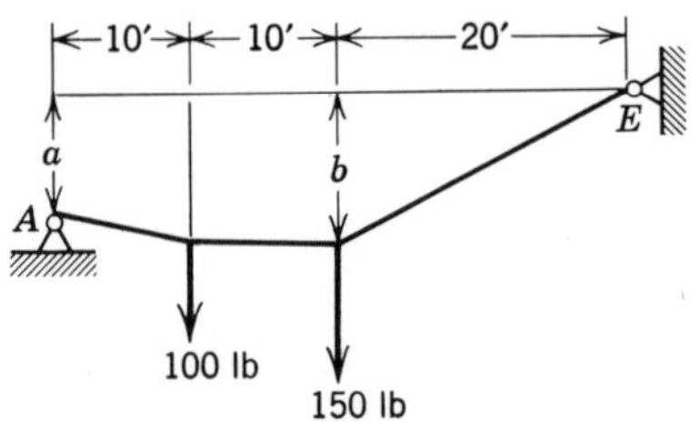

Figure P7.9

(11) The two cables of a suspension bridge with a span of 4000 ft and a sag of 450 ft support a uniform horizontal load of 40 $\times$ 10^3 lb/ft. Compute the maximum tension in each cable and the angle θ between the horizontal and the cables at the tower supports (located at equal elevations).

(12) The light flexible cable in Fig. P7.10 supports a load of 150 lb/ft uniformly distributed along the horizontal. Find the minimum tension, $T_{\min}$, and the maximum tension, $T_{\max}$.

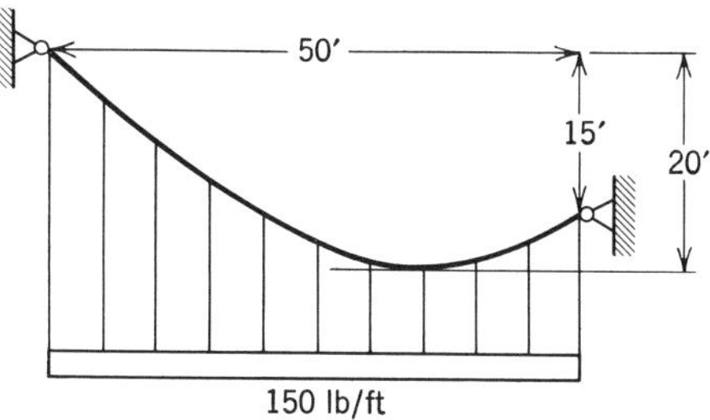

Figure P7.10

Answers to Problems

Chapter 7

(1) Method of Sections

$$F_{BC} = \frac{4P}{7\sqrt{3}} \text{ compression}$$

(2) a. $F_{AB} = .66$ lb. T
$F_{AD} = 9.6$ lb. $T = F_{DC}$
$F_{BD} = 5$ lb. T
$F_{BC} = 11$ lb. C
b. Yes

(3) $F_A = +2W$
$M_A = 4W\ell$

(4) $M_E = \dfrac{W\ell}{6\sqrt{3}}$

(5) $F_{BA} = 2\sqrt{3}\,P$ tension

$$F_{BC} = \frac{P}{\sqrt{3}} \text{ tension}$$

(6) $R_L = 0\bar{\mathbf{i}} + 1750\,\bar{\mathbf{j}}$ lb
$R_R = 1250\bar{\mathbf{j}}$ lb

(7) –

(8) $\overline{\mathbf{R}} = 0\overline{\mathbf{i}} + 500\overline{\mathbf{j}}$ lb
$\overline{\mathbf{M}} = +8750\overline{\mathbf{k}}$ ft-lb
(9) $\overline{\mathbf{R}}_L = 0\overline{\mathbf{i}} + 166\overline{\mathbf{j}}$ lb
$\overline{\mathbf{R}}_R = 84\overline{\mathbf{j}}$ lb
(10) $E_x = 417$ lb, $E_y = 259$ lb
(11) $\theta = 8.5°$, $T_{max} = 2.7 \times 10^8$ lb
(12) $T_{min} = 4167$ lb
$T_{max} = 6480$ lb

chapter 8

Friction

Objectives

Here we seek to provide an understanding of the nature of dry friction and to provide a method for solving equilibrium problems involving friction. This chapter should help the reader be able to:

1. Define dry friction and describe the key features of the behavior of friction forces.
2. Define and explain the concepts coefficient of friction and angle of friction.
3. State the three types of friction problems and correctly classify typical problems.
4. State and apply correctly the general procedure for solving simple problems of each type.
5. Derive and use properly the formulas describing the behavior of square threaded screws with friction.
6. Derive the relation $T_L = T_S e^{\mu\theta}$ for flat belts with friction and use it to solve simple problems.

Frictional effects, which represent an essential aspect of the world around us, are very common. In general, frictional effects are associated with resistance to relative motion. A few of the different types of friction are internal friction in materials, rolling friction, fluid friction, and dry, sliding friction. Here we will limit our study to dry, sliding friction, also called Coulomb friction, which is defined as the tendency of dry surfaces that are pressed together by a normal force to resist motion relative to each other.

The force that resists or opposes relative motion is called the friction force and is tangent to the contact surface.

Our first task is to develop an understanding of some characteristics of dry friction. One important characteristic is that *friction forces are always directed so as to resist the relative motion of the two surfaces in contact.* Consider Fig. 8.1. Here, block *B*

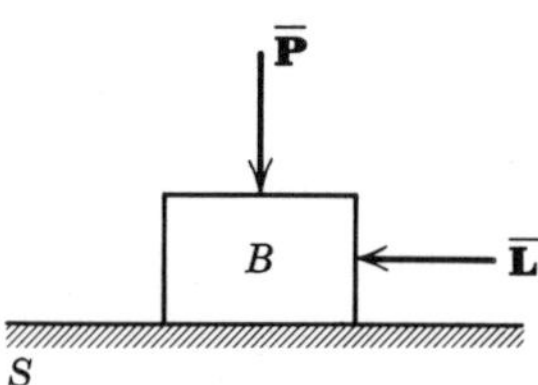

Figure 8.1

is pressed against a flat surface *S* by a force $\bar{\mathbf{P}}$ (considered here to include the weight of block *B*) and is subjected to a lateral force $\bar{\mathbf{L}}$ parallel to the surface *S*. You have already learned that a first step in almost any statics problem is to draw a ____ ____ ______.

free body diagram

Therefore, sketch a free body diagram of block *B*, showing $\bar{\mathbf{P}}$ and $\bar{\mathbf{L}}$, in the margin.

Now a decision must be made regarding representation of the resultant force acting between the plane surface *S* and block *B*. It is convenient to divide this force into components normal and tangential to the plane of contact. With the block in static equilibrium, we expect the normal component $\bar{\mathbf{N}}$ acting on *B* to be equal in magnitude to the force __ and to be directed positive ______________.

upward/downward

$\bar{\mathbf{P}}$

upward

Add this force to the free body diagram. Now consider the tangential component, which is the friction force $\bar{\mathbf{f}}$. The direction of the friction force is determined by the requirement that this force must oppose relative motion of the block and surface. Therefore the friction force should be directed positive to the ________.

left/right

right

Add this force to your sketch and compare your

diagram with the completed free body diagram in Fig. 8.2.

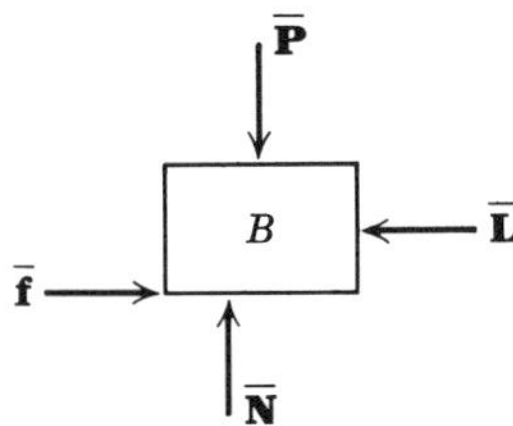

Figure 8.2

Now let us consider the magnitude of the force $\bar{\mathbf{f}}$. From experience we know that if the block is in equilibrium and the force $\bar{\mathbf{L}}$ is small compared to the normal force pressing the surfaces together, then the magnitude of the friction force is ________ ____ L.
equal to/different from

equal to

Thus, if L equals one pound, f also equals one pound and if L were equal to 17.3 pounds, we would expect f to equal ___ lb.

17.3

From experience we also know that there is a maximum value of L, with N constant, for which the block will remain in equilibrium and not slip. The second characteristic of dry friction is a *limiting, maximum value of f associated with a given normal force N*. Another experimentally proven fact is that the ratio of the maximum friction force, $f_{\max}$, to N usually can be considered a constant for a given pair of materials of fixed surface condition. We define this proportionality constant as $\mu = \dfrac{f_{\max}}{N}$. The letter μ is called the coefficient of static friction and is considered a constant that depends only on the materials in contact and their surface finish. Thus, if we know μ and N we may determine $f_{\max}$, since $f_{\max} =$ ___.

μN

What happens when L, the lateral force, exceeds $f_{\max}$? The answer is that block B begins to slip on the plane and we now have a problem of dynamics rather than statics.

When motion is taking place, *kinetic* friction is involved. Experiments have shown that the coefficient of kinetic friction, defined as $\mu_k = f_{\text{kinetic}}/N$, is relatively independent of velocity and somewhat less than the coefficient of static friction.

Let us summarize what has been presented. First, with friction present and with a relatively small lateral force, the friction force f is equal in magnitude to the ____ ____ ____.

lateral force − L

There exists a maximum possible friction force that can be found from the definition of the friction coefficient as $f_{\text{max}} =$ ____.

μN

We know, furthermore, that this maximum value of friction force occurs only when relative motion, or slip, impends. We thus conclude that for motion not impending, the friction force equals the lateral force; for motion impending, $f_{\text{max}} =$ ____; and when motion is taking place (a dynamic situation), the friction force is equal to $\mu_k N$ where μ_k is the coefficient of ____ friction.

μN

kinetic

To emphasize that friction forces always oppose relative motion, consider the problem of a block B whose *top* view is shown in Fig. 8.3. Block B weighs

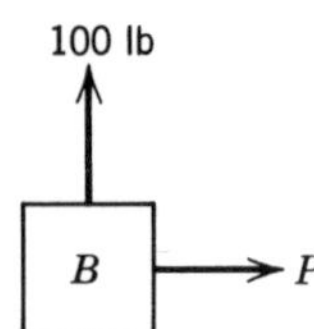

Figure 8.3

500 lb. and is resting on a plane surface. The coefficient of static friction between the block and the surface is 0.30. The question is: what value of the force P acting in addition to the 100-lb force will cause the block to slip on the plane? The normal force between block B and the plane (not shown) has a magnitude N equal to ____ lb.

500

Therefore, slip will occur when the resultant friction force has a magnitude $f = \mu N$ equal to ____ lb.

150

Thus, slip will occur when the resultant external lateral force reaches a magnitude of 150 lb. This resultant lateral force is the *vector sum* of the 100-lb force and $\overline{\mathbf{P}}$. Make the calculation necessary to show that the minimum value of P necessary for slip to occur is ______________ lb.

$\sqrt{150^2 - 100^2} = 112$

It is useful to introduce the concept of angle of friction, defined as the angle, when slip impends, between the *resultant* contact force, $\overline{\mathbf{R}}$, and the normal to the surface, $\overline{\mathbf{N}}$. In Fig. 8.4 this angle is called ___.

ϕ

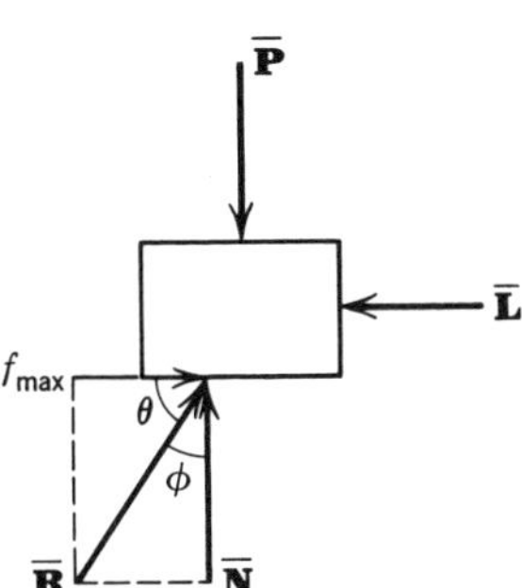

Figure 8.4

We also see that the tangent of this angle equals f_{max} divided by ___.

N

From the relations $\tan \phi = f_{max}/N$ and $\mu = f_{max}/N$, we conclude that the $\tan \phi$ equals ___.

μ

Alternatively, we conclude that $\tan^{-1}\mu$ equals ______________.

ϕ, the angle of friction

We now examine the relationship of contact area to frictional behavior. A basic approach to dry friction begins with an experimentally derived relation between the maximum friction force on a small plane element of contact area and the normal force on that area element, that is, $f_{max} = \mu N$. Experiments show this relationship to hold regardless of the size of the area involved. Furthermore, we note that the normal force and the friction force need not be distributed uniformly over the contact area.

To determine the behavior of an entire system, we must consider the sum of the contributions of the small elements, each of which obeys the relation $f_{max} = \mu N$.

Let us call $d\overline{\mathbf{f}}$ the friction force on any element of area dA and $d\bar{\mathbf{N}}$ the normal force acting on that area. If relative motion, or slip, impends then for every dA,

μdN df_{max} equals ___.

In the general case we use the previous result and determine the system behavior by summation (or integration) of $d\bar{\mathbf{f}}$ and $d\bar{\mathbf{N}}$ over the entire contact area. Such general problems are often quite involved and are beyond the scope of this book. However, simplification can be achieved when surfaces are plane and only translatory motion occurs.

For example, consider the block B of Figs. 8.1 and 8.2. First, if motion impends, we know that for every element of area $df_{max} = \mu dN$. Furthermore, we know that all elements of the normal force $d\bar{\mathbf{N}}$ are parallel and all elements of the friction force, $d\bar{\mathbf{f}}$, are parallel. Thus the sum of the equations of the form df_{max}

μN $= \mu dN$ is the $\Sigma_{area}\, df_{max} = f_{max} = \Sigma_{area}\, \mu dN =$ ___.

However, knowledge of the resultant friction force and normal force is not sufficient for the determination of the exact *distribution* of these forces. In fact, we *cannot* determine with the laws of statics alone what these distributions are. We may, however, find the *location* of the resultant forces $\bar{\mathbf{f}}$ and $\bar{\mathbf{N}}$ using equations of static equilibrium. Consider moment equilibrium of block B in Fig. 8.2. Clearly, the re-

cannot sultant normal force $\bar{\mathbf{N}}$ ________ be colinear with

can/cannot

the force $\bar{\mathbf{P}}$.

For moment equilibrium to be maintained, $\bar{\mathbf{N}}$ must be located so that the distance between the forces $\bar{\mathbf{N}}$ and $\bar{\mathbf{P}}$, multiplied by N or P, will equal the distance

$\bar{\mathbf{f}}$ $\bar{\mathbf{L}}$ between the forces __ and __ multiplied by f or L.

Thus the location of the resultant normal force $\bar{\mathbf{N}}$ has been found by moment considerations even though the exact distribution of $d\bar{\mathbf{N}}$ cannot be determined.

We conclude that the laws of statics afford, at most, the determination of the magnitude and location of the *resultant* friction and normal forces but do not permit computation of the force distributions.

We shall now consider a general approach to the solution of problems involving friction. Most problems involving friction can be classified as one of the following three types:

1. Motion impends or is taking place at a given number of *known* contact surfaces.
2. Impending or actual motion exists but the exact nature of this motion and the surfaces on which the motion takes place are not known.
3. Whether motion impends, or is taking place is not known.

We will be referring to these three types of friction problems throughout this chapter, so be sure to learn them before continuing.

Since the proper approach to problems depends on the type involved, you must first *correctly* identify the type of problem at hand. To do this you need answer two questions. The first question is: "Are you sure motion impends or exists?" If the answer is "no" then you have a problem of type __. 3

If your answer is "yes," then the problem is either of type __ or type __. 1 2

If your answer was "yes," you must answer a second question. "Do you know *definitely* for every surface whether or not motion impends or is taking place?" If you know the surfaces at which motion is taking place, then the problem is of type __. 1

If you do not know, then the problem is of type __. 2

Let us now investigate the methods of solution of these three types of problems by examples. First, consider the missile silo door shown in Fig. 8.5. During installation, this 8000-lb door must be propped in position by bar *BC*. Suppose you are asked to deter-

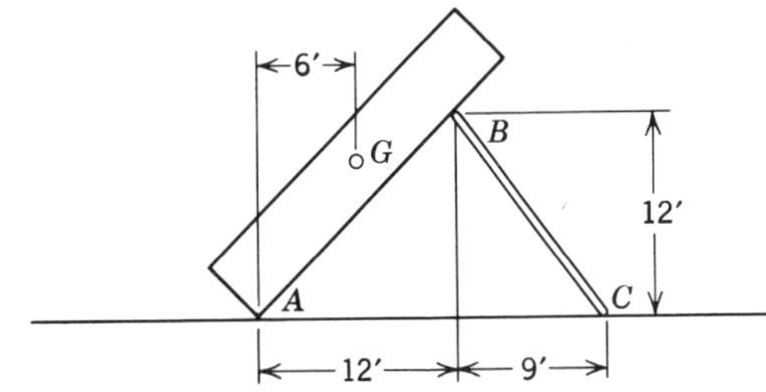

Figure 8.5

mine the minimum allowable coefficient of friction between the support BC and the surface at point C so that slip does not occur, that is, for the system to be in equilibrium. In other words, you are being asked for the coefficient of static friction for which motion impends between the surfaces in contact at point __.

C

Also, you can conclude that there is no relative motion taking place at points A or B, since we have assumed the condition of ________.

equilibrium

Since you know that motion impends and the exact nature of the situation at all surfaces, this problem is of type __.

1

The general procedure for problems of type 1 is to write the equation $f = \mu N$ *for all surfaces where relative motion impends and then to solve the problem using the laws of equilibrium.* Clearly, it is essential to assign the correct direction to the friction force. This direction must be assigned so as to ___________ relative motion between the surfaces
aid/oppose
at C.

oppose

In most problems of equilibrium it is convenient to draw a free body diagram. Therefore, draw a free body diagram of the entire system, that is, a diagram of the door and the bar together, in the space provided. Name the reaction components at A, $A_x(\rightarrow)$ and $A_y(\uparrow)$, and those at C, $f_c(\leftarrow)$ and $N_c(\uparrow)$. Notice that the force components at B are disregarded, since they are considered internal to our system.

The laws of statics applied to this system will yield four independent equations—three equilibrium equations and a friction equation at C. However, these equations will contain five unknowns—A_x, A_y, f_c, N_c, and μ. Thus we can not solve for the unknowns of the system as we have chosen it. Our alternative is to consider the door and the bar as individual ________ (or subsystems) of our system and to draw a ____ ____ ______ of each of them. The force components at B, B_x, and B_y, are now external to the door and the bar, and must be included. Sketch the free body diagrams in the space provided.

components
free body
diagram

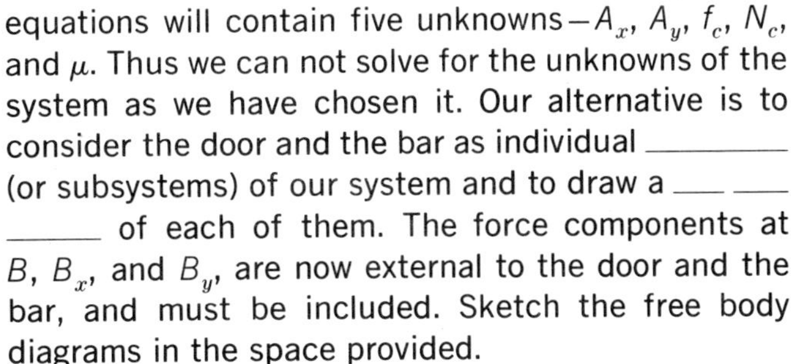

The positive force directions chosen and used subsequently in equations are shown in Fig. 8.6. Make sure your free body diagrams agree with these.

Now with the seven unknowns (A_x, A_y, B_x, B_y, f_c, N_c, and μ), we have six independent equilibrium equations and the friction equation. We now seek a way of solving for $\mu_{\min}$. In this case there is a simple approach that involves writing a single equation containing the unknowns f_c and N_c only. Considering member BC, we can obtain this equation, since the sum of the _______ of forces about point __ must equal zero.

moments B

The sum of moments about point B yields the equation ______________.

$-12f_c + 9N_c = 0$

This equation can be solved for the ratio f_c/N_c, that is, $f_c/N_c =$ ______________.

$9/12 = 3/4 = 0.75$

Since by definition $\mu = f_c/N_c$, when slip impends, we have solved the problem. The minimum value of μ required for equilibrium is $\mu_{\min} =$ ____.

0.75

Let us now look at *a different problem* involving the same system shown in Fig. 8.5. We ask the following question. If the silo door is supported as shown and if the coefficient of friction between the contact

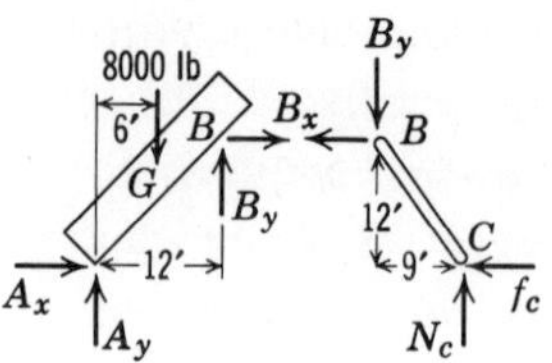

Figure 8.6

surfaces at points A and B equals 0.5 and at point C equals 0.8, is the system in equilibrium? In this case do we know if motion impends or is taking place?
No ______. Therefore, we classify this problem as type
Yes/No
3 __.

Since in problems of type 3 *we do not know if equilibrium exists, it is necessary to make an assumption*. Ordinarily, the best approach is to assume equilibrium exists, then solve the problem and check to see if any friction conditions have been violated. Again it is convenient to split this system into two components and to draw two free body diagrams. Naming the reaction forces at C, $C_x(\rightarrow)$, and $C_y(\uparrow)$, draw an appropriate pair of free body diagrams (in the space provided).

It is important to note a significant change in the situation at C between this example and the previous example. Since we do not know if slip impends at C,
is not the relation $C_x/C_y = \mu$ ________ necessarily true.
is/is not

A pair of free body diagrams with the notation

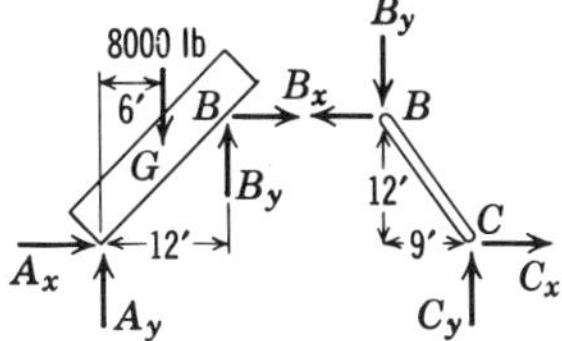

Figure 8.7

to be used in equations is shown in Fig. 8.7. Now we must write a set of equilibrium equations to be used to solve for the six unknowns of Fig. 8.7. Let us begin by writing equilibrium equations for the bar *BC*. Write the equations for the sum of the forces in the *x* and *y* direction, and the sum of moments about point *C* (in the space below).

$C_x - B_x = 0$ (1)
$C_y - B_y = 0$ (2)
$12B_x + 9B_y = 0$ (3)

For the door, write equations representing the sum of the forces in the *x* and *y* directions, and the sum of moments about point *A*.

$A_x + B_x = 0$ (4)
$A_y + B_y - 8000 = 0$ (5)
$12\,B_y - 12\,B_x - 48{,}000 = 0$ (6)

Now on scratch paper solve these equations to find $A_x =$ ________, $A_y =$ ________, $B_x =$ ________, $B_y =$ ________, $C_x =$ ________, and $C_y =$ ________.

$A_x =$ 1715 lb
$A_y =$ 5715 lb
$B_x =$ −1715 lb
$B_y =$ 2285 lb
$C_x =$ −1715 lb
$C_y =$ 2285 lb

The negative signs associated with B_x and C_x

wrong

mean that the ________ direction was assumed
right/wrong
for these forces.

less

Our calculations were made with the assumption that the system is in equilibrium. This means that the ratio of the tangential force to the normal force at any potential sliding contact surface must be ________ than the coefficient of friction at that
greater/less
point.

μ

To check our assumption of equilibrium we must now show that we have not violated any friction conditions. That is, we must be certain that nowhere have we required f/N to be greater than __.

A_x

A_y

Is the friction condition violated for the contact surface at point A? Here we associate the friction force with the force component __ and the normal force with the force component __.

A_x A_y

$\mu = 0.5$

Thus our check is that the ratio of __ to __ must be less than or equal to ____.

is not

Dividing A_x by A_y shows that the friction condition ________ violated at point A.
is/is not

C_x C_y $\mu = 0.8$

Although slip does not impend at point A, it may impend at some other point. Therefore we must check every point at which slip might occur. Next consider point C. Here we have the requirement that the ratio of __ to __ must be less than ____.

less

is not

If we take the ratio of the values of C_x and C_y obtained previously, we find it ________ than 0.8
less/greater
and, therefore, motion ________ impending or taking
is/is not
place at point C.

2830 lb 403 lb

A final check required is at point B. Here we have a complication, since the components B_x and B_y are not normal and tangential to the surface on which slip might occur. To check the friction condition at B, we must first transform B_x and B_y into B_{normal} and $B_{\text{tangential}}$. So calculate them to find that B_{normal} = ____ and $B_{\text{tangential}}$ = ____.

less

Now taking the ratio of $B_{\text{tangential}}$ to B_{normal} we find it ________ than the coefficient of friction at that
less/greater

point and, therefore, the friction requirement ________ violated at *B*. is not
is/is not

We have found that the friction forces required at each point are less than the maximum allowable friction forces for equilibrium at that point. Therefore the friction condition is not violated at any of these points, and we conclude that there is sufficient friction present for the system to be in equilibrium as was assumed. Remember that you must check *every* point where slip might occur.

Suppose the friction force required at some surface had been greater than the product μN. Since this is impossible, we would conclude that the system ____________ in equilibrium and, therefore, the was not
was/was not
problem is of _______ and not of statics. dynamics

Let us examine another type of problem. Consider the two-stage rocket (shown in Fig. 8.8) that is

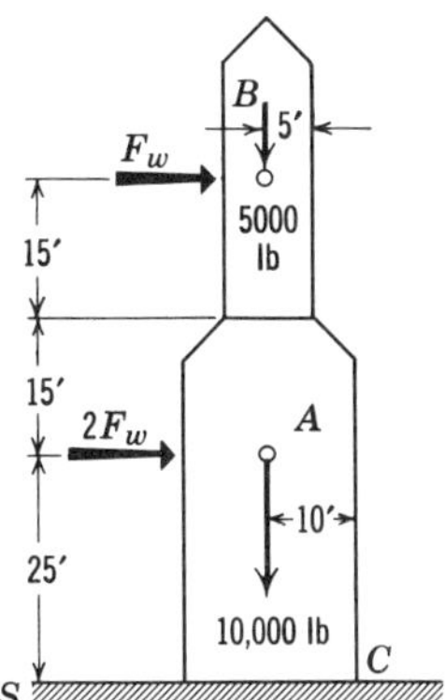

Figure 8.8

being assembled. At present, stages *A* and *B* are unattached; *B* is resting on top of stage *A*. Stage *A* rests on the ground surface *S*. We know the coefficient of friction between all surfaces is 0.4. The question is: "If wind exerts a force F_w on stage *B* and a force $2F_w$ on stage *A*, what is the largest value of F_w this rocket can withstand and remain in equilibrium?" Stage *A* weighs 10,000 lb and stage *B* weighs 5000 lb.

In this problem we are asked for the limiting value of F_w so that equilibrium is maintained. We do not know, however, in what way motion might occur. Since we seek impending motion with the type of motion unspecified, this is a friction problem of
2 type __.

Let us now consider ways motion might take place. First, slip might occur between stages *A* and *B* or between stage *A* and *S*. Also, stage *B* might tip over or stages *A* and *B* might tip together. Thus, motion might occur in any of four ways or any combination of them. We do not know which in fact will occur. A recommended approach for this type of problem is to look carefully at the problem and choose what you think is the *most probable* mode of motion. Then, assume that motion occurs in this manner and solve for the forces required. Finally, make certain that no conditions of equilibrium or friction have been violated anywhere in the problem. If you find no violations, then you have selected the correct mode of motion! On the other hand, if there are violations of equilibrium or friction conditions inherent, then the
wrong answer determined is __________.
correct/wrong

However, in the latter case you can use the results from the first attempt to make a better assumption. Likely, you will have the correct mode of motion for this second time through. Let us now use this approach on the subject problem. We choose as our first assumption that, for the entire rocket, stages *A* and *B* together, tipping impends about point *C* on *S*. For this assumption, draw in the space provided an appropriate free body diagram considering the remainder of the system to be in equilibrium (at present, omit the normal and friction forces acting on the base).

We now consider where and how to draw the resultant friction and normal forces acting at the base. First, since any sliding that occurs will be parallel to the surface *S*, we know that the friction force must be parallel to this surface. We also know that this
oppose friction force must be directed so as to __________
aid/oppose
relative motion of the contacting surfaces. We there-

fore choose the friction force $\bar{\mathbf{f}}$ to be a horizontal force directed positive to the ________. left
right/left

Draw the friction force $\bar{\mathbf{f}}$ on your free body diagram. Consider next where the resultant normal force $\bar{\mathbf{N}}$ should be drawn. Certainly, $\bar{\mathbf{N}}$ must act on that part of the base in contact with S. Also, as the wind force increases and the rocket begins to tip, the side of the base toward the wind will tend to rise while the side away from the wind will be pressed more *firmly* against S. Thus as F_w increases, the resultant force $\bar{\mathbf{N}}$ will move ______________ the point C. toward
away from/toward

It is assumed that the exact distribution of forces will be such that moment equilibrium is maintained. Thus, as the tilting moment of the wind on the rocket system is increased, we expect $\bar{\mathbf{N}}$ to move away from the center of the rocket and toward C in an attempt to maintain _____ ________. moment equilibrium

The limiting case and the maximum possible distance that $\bar{\mathbf{N}}$ can move is to the point __. C

This is reasonable since, upon tipping, point C will be the only point in contact with the surface. Therefore, we conclude that if tipping impends, we must draw the normal force $\bar{\mathbf{N}}$ in our free body diagram as acting through point __. Add $\bar{\mathbf{N}}$ to the free body diagram. C

Compare your free body diagram with Fig. 8.9. We now solve this problem for the magnitude of F_w and then check to see if any conditions of equilibrium or friction have been violated. First we note that in this problem there are ________ unknowns. three (F_w, f, N)

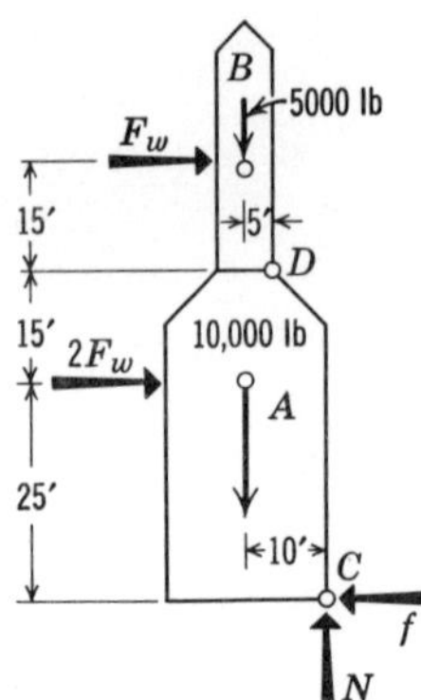

Figure 8.9

three — Since this is a plain problem and there are ____
equilibrium equations available, can we solve for the
Yes — unknowns? ______.
Yes/No

Since we have assumed tipping about C with no
need not — slipping, the relation $f = \mu N$ ____________ be true
must/need not
here.

To solve the problem, let us investigate the lines
of action of various forces. Two of the forces are con-
C — current at point __; therefore, a moment equation
C F_w — about point __ involves only one unknown, namely __.

Write the equation for the moment of forces
about point C in the space provided. Solve this equa-
1430 lb — tion to find that F_w = ______.

This is a tentative value for the maximum wind
force F_w. We must determine now that no conditions
of equilibrium are violated.

To do this, use force equilibrium to find the two
15,000 lb — remaining unknown forces. Thus N = ______ and
4290 lb — f = ______.

We now check for slipping at point C. Since the
less have not — ratio f/N is __________ than 0.4, we ____________
greater/less have/have not
violated the friction condition at point C.

Next we must check for the motion of stage *B* relative to stage *A*. First, draw a free body diagram of stage *B* alone in the space provided. Include a friction force f_D and a normal force N_D at the appropriate place, considering that tipping as well as slipping may impend.

With the equations of force equilibrium for this free body, you can solve for f_D and N_D, as f_D = ______ and N_D = ______. 1430 lb 5000 lb

Now we check and find that since f_D/N_D is ____________ (greater/less) than 0.4, slipping at this surface ____________ (will/will not) occur. less will not

The final check is to see if the upper stage will, in fact, tip about point *D*. This is accomplished by summing moments about *D*, assuming that tipping impends. Use the space below to write an equation for the sum of the moments about point *D* and solve it for $F_{w-\text{tip}D}$ = ______. 1670 lb

Since the wind force required for tipping at *D* is greater than that for tipping at *C*, we conclude that the system would more readily tip about point *C* as was originally assumed. Since we have now checked all possible modes of motion and found no violations

of equilibrium or friction requirements, we conclude
was correct that our orignal solution ________________.
was correct/needs revision

Of course, had *any* equilibrium or friction requirement been violated, our original answer would have been meaningless and another solution attempt would have been necessary. This second attempt of course, based on the results of the first, would include assumption of a new mode of motion, determination of the resulting force-magnitude requirements, and
equilibrium friction checking with all ________ and ______ requirements.

This process would then be repeated until a solution satisfying all equilibrium and friction requirements is found.

An alternative approach to this type of problem is to assume, one at a time, that slipping or tipping occurs in each possible mode. The value of F_w required for each mode is then calculated. The lowest value indicates the actual way that slipping or tipping would occur and the desired maximum size of F_w for equilibrium.

Our next topic will be the influence of friction on several types of machine elements. Figure 8.10 shows

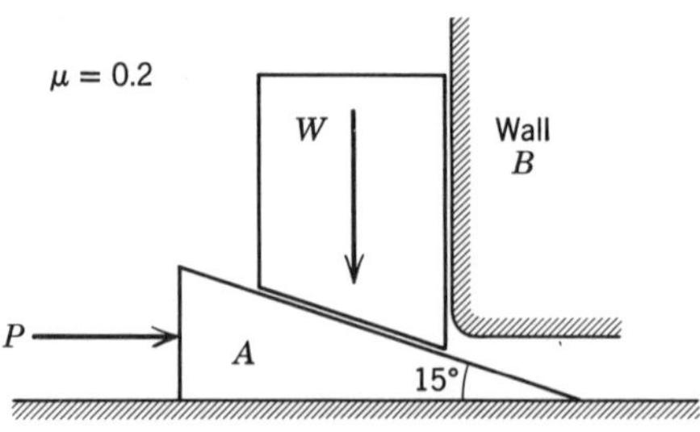

Figure 8.10

a wedge *A* being used to lift a 1000-lb weight *W*. For this problem, the coefficient of friction is $\mu = 0.2$ for all surfaces that have sliding contact. We seek the minimum force *P* required to raise weight *W*. Since we know that slip impends at all surfaces we can
1 classify the problem as type __.

The proper procedure is thus to write $f = \mu N$ at all sliding surfaces, to draw free body diagrams of appropriate elements, and to write and solve associated equilibrium equations.

As your first element, select the wedge *A* and draw its free body diagram in the space provided. In

your diagram, draw forces as normal and tangential components where appropriate. Next compare your diagram with Fig. 8.11, considering, particularly, the directions chosen for the tantential friction forces. If there is any question, review the basic nature of friction described at the beginning of this chapter.

Now examine the free body diagram carefully. Since there are no locations of forces given, we cannot use the equilibrium requirement that the sum of the ______ of forces about any point is zero. moments

We have remaining two equations of force equilibrium and two friction equations. Since there are ________ unknowns, we ___________ determine all of five cannot
(number) can/cannot
these unknowns with these equations alone.

So, draw another free body diagram for the weight *W*, in the space provided. Your diagram should agree with Fig. 8.12.

Here, we find that a moment equilibrium equation ___________ be written and that there are cannot
can/cannot
___ useful equilibrium equations plus ___ friction two two
equations.

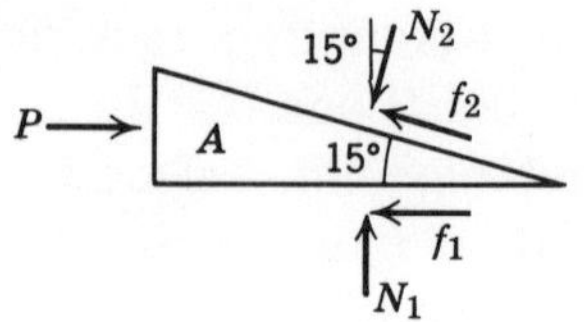

Figure 8.11

four Since W weighs 1000 lb, there are only ___ un-
can known quantities and we ___________ determine
can/cannot
them from the equations available.

Next (on scratch paper) write the appropriate

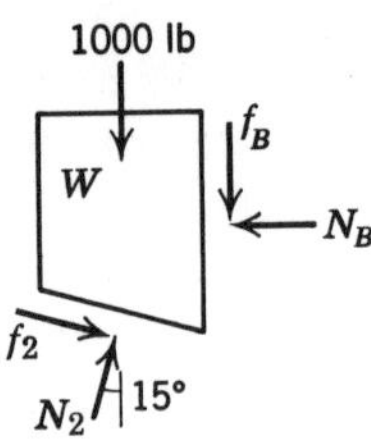

Figure 8.12

equations of force equilibrium and friction, and solve
1213 lb 243 lb them to find that N_2 = _____ and f_2 = _____ (no need to solve for f_B and N_B at wall).

With f_2 and N_2 known, write and solve the equilibrium and friction equations for wedge A to determine the remaining unknown forces. In particular,
771 lb solve to show that the force P equals _____.

As an exercise, solve the following problem. If P equals zero pounds, what is the minimum coefficient of friction $\mu_{\min}$ that will prevent the wedge from slipping from under the weight W? This problem is solved in the same manner as the previous one with the exception that friction forces have different directions, since a different mode of relative motion is involved. If you solve this problem you will find that
0.135 $\mu_{\min}$ equals ___.

Another common machine element, the square threaded screw, can be readily treated as a special case of a wedge or an inclined plane. Figure 8.13 shows a block N (corresponding to the nut on the

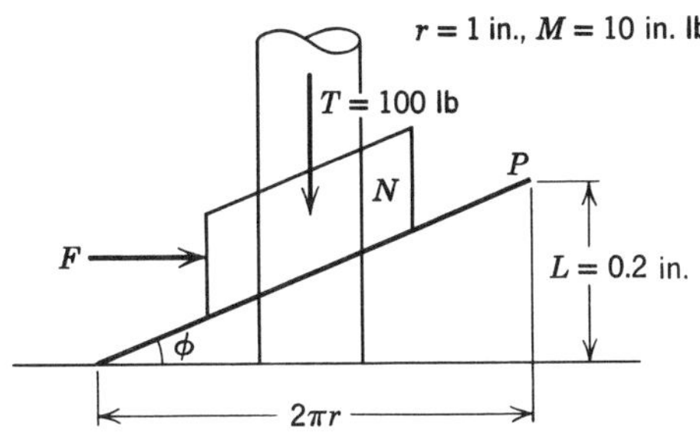

Figure 8.13

screw) being pushed up an inclined plane P (which corresponds to the screw threads) by a force F. *We may think of this situation as representing a screw by simply wrapping the plane and nut around the screw axis.* The thrust T of the screw/nut combination corresponds to the weight of a block on a plane. The angle of the plane, ϕ, is obtained by considering its tangent as $L/2\pi r$ where r is the radial distance to the center of the thread and L is the lead of the screw. Ordinarily a moment M is used to twist a nut on a screw. This moment M produces the force F at a distance r from the axis, that is, $M = rF$.

With this brief introduction we will now examine the following specific problem involving a square threaded screw. The problem is: for a screw with an average thread radius (r) equal to 1 inch and a lead (L) of 0.2 inches per revolution, what is the maximum coefficient of friction that will allow a moment of ten inch-pounds to produce an axial thrust T equal to 100 pounds?

As usual, our first step in solving the problem is to draw a free body diagram. Whether we can obtain a solution and how much effort will be involved depends upon a proper choice for the free body diagram. A good choice here is to draw a section of the nut, similar to that labeled N in Fig. 8.13. Do this, being careful to determine the direction of your friction force properly.

To check your results, compare with Fig. 8.14.

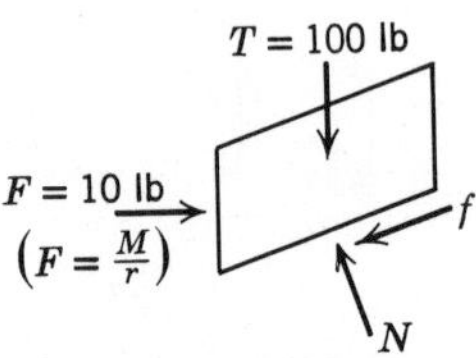

Figure 8.14

From the problem statement, it is understood that a limiting value of μ is sought, since motion impends. Thus with the type of problem known (type 1), consider the nut in equilibrium and treat it accordingly. We may calculate F, since M and r are known;

M r 10 — $F =$ __/__ $=$ __ lb.

Also, we can calculate the angle of inclination

$2\pi r$ — of the plane from the relation: tangent $\phi = 0.2/$____.

Then we write and solve the two force equilibrium equations for the two unknown forces f and N. Do

6.8 lb 100 lb — this to find that $f =$ _____ and $N =$ _____.

You are ready to determine the maximum possi-

0.068 — ble coefficient of friction as $\mu_{max} = f/N =$ ____.

Other problems for square threaded screws are treated in a similar way with the same equations. Other thread profiles can also be treated, but their analysis is more involved because of additional moments that are present.

A somewhat different example of friction is involved when a rope or a belt slides on a curved surface. Consider the case shown in Fig. 8.15—a belt with tension T_{large} at one end and T_{small} at the other for which slip impends on the fixed surface S. To write the relationships describing this problem, we consider the equilibrium of a small elemental section of the belt such as the one enclosed in the dotted circle in Fig. 8.15. If a free body diagram of this ele-

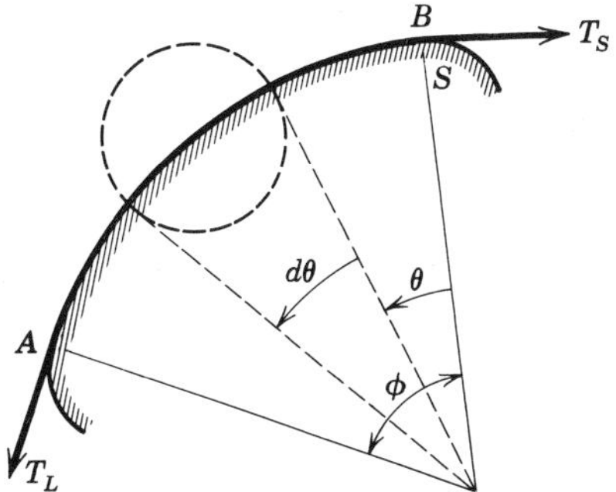

Figure 8.15

ment is drawn as in Fig. 8.16, the equations of equilibrium can be used to develop a differential equation which, when integrated along S, will describe the behavior of the flat belt as a whole.

First, let us explain why we drew the free body diagram as shown. In particular, what should be the direction of the friction force df in Fig. 8.16? Since

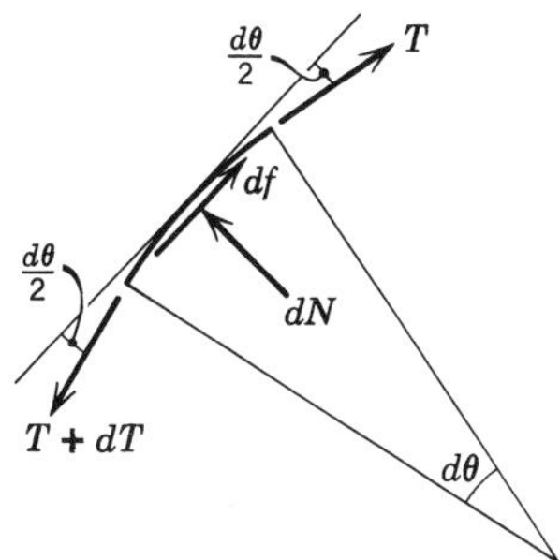

Figure 8.16

the larger force is pulling the belt downward and to the left, any friction force on an element would be directed so as to oppose this motion and thus would be directed positive ________________ and to the
upward/downward

__________.
left/right

upward

right

For the same reason, the magnitude of the tensile force of the belt as we move from the top right to the lower left of S in Fig. 8.15 must gradually

increase ________________.

increase/decrease

Therefore, the tension at the lower left of the

plus — curved element in Fig. 8.16 is T __________ an incre-

plus/minus

ment dT indicating that tension increases in that direction.

The radius of curvature of the belt at this point and the angle $d\theta$ describe the length of the elemental segment of belt. N, the normal force, must be positive (compressive) between the belt and surface. Let us now apply the equations of force equilibrium in the normal and tangential directions to this element. First, equilibrium of forces in the normal direction

$d\theta/2$ $\quad T \sin (d\theta/2)$ — yields the equation: $dN - (T + dT) \sin$ ____ $-$ ________ $= 0$.

Similarly, tangential force equilibrium yields the

$T \cos (d\theta/2)$ — equation $(T + dT) \cos (d\theta/2) - df -$ ________ $= 0$.

Since the angles $d\theta$ and $d\theta/2$ are small and for small angles $\sin \alpha$ approximately equals α while $\cos \alpha$ approximately equals 1, we can simplify the two equations to the forms ____________________ and ______________.

$dN - (T + dT)(d\theta/2) - T(d\theta/2) = 0$

$T + dT - df - T = 0$

If we use the fact that the product of differentials is negligibly small compared to the first power of these differentials, we can further simplify the first equation to __________ and the second equation to ________.

$dN - T\,d\theta = 0$

$dT - df = 0$

Since we have specified that slip impends, we know that $\mu\, dN = df$; hence the first equation can be written as $dN = df/\mu = T\,d\theta$. If we now use the second equation to replace df by dT in the first, our resulting

$dT/\mu = T\,d\theta$ — equation is ________. If this equation is written as $dT/T = \mu\, d\theta$, then we have a differential equation in which the variables are separated. Before we integrate this equation, let us investigate certain characteristics of our derivation. In Fig. 8.15 the forces T_L and T_s are the limits on the integral of dT/T. They are independent of the shape of surface S. Furthermore, the limits on the integral of $d\theta$ are $\theta = 0$ and $\theta = \phi$ where ϕ is simply the maximum angle turned by the tangent (and normal) to the belt. We thus conclude that our differential equation holds for any curved surface S with which the belt remains in contact, and that it is only the total _____ turned and not the exact _____ of the surface which is significant.

angle
shape

Now integrate this expression to obtain the relationship

$$\ln T \Big|_{T_s}^{T_L} = \mu\theta \Big|_0^{\phi}.$$

This relationship can also be written $\ln T_L - \ln T_s = \mu\phi$. According to the laws of natural logarithms we may write $\ln (T_L/T_s) = \mu\phi$. Usually it is convenient to write this in an exponent form as $T_L = T_s \exp$ (_____).

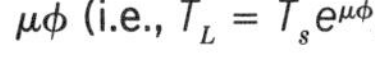

This final equation states that the large tension equals the small tension multiplied by exp ($\mu\phi$). As a review let us solve the following problem. Figure 8.17 shows a 100-pound weight W attached to a

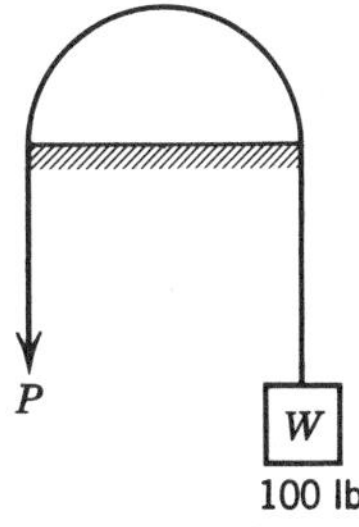

Figure 8.17

rope that passes over a stationary drum and has a force P acting on its other end. The coefficient of friction between rope and drum is 0.3. We ask: what is the maximum and minimum force P for which the system will be in equilibrium? First, if we seek the *maximum* force, P_{max}, we know that motion impends such that the weight W will move _____________.
downward/upward

upward

For this case we know that the force P_{max} must be _________ than 100 lb and, in the belt friction equa-
greater/less
tion, we associate T_L with the force ___ and T_s with _______.

greater

P_{max}

$W = 100$ lb

A substitution of these values into the equation yields the relationship ________.

$P_{max} = 100\, e^{\mu\phi}$

Here ϕ is the angle of contact measured in radians. For this circular surface ϕ equals ______. We write the final relation as __________.

π radians

$P_{max} = 100\, e^{0.3\pi}$

With your slide rule, or log table, solve the equation to find that P_{max} = ____.

256 lb

For the minimum value of P we must reconsider the problem. This minimum value, P_{min}, is associated with impending motion of W – ______________.
upward/downward

downward

In this case, we associate T_L with ________ and T_s with ____.

$W = 100$ lb

Since we have the same angle of contact as the previous case, we can write our final equation as __________.

$100 = P_{min} e^{0.3\pi}$

Solve this equation to find P_{min} = ____.

39 lb

These problems are simple so long as you remember that impending motion always takes place in the direction of the maximum tension.

This brings to a close our discussion of friction. In essence this chapter has said that with friction problems you must bear in mind two characteristics of friction. First, friction forces are always directed so as to ________ relative motion. Second, the fric-
aid/oppose
tion law, $f = \mu N$, applies only if relative motion ______ or __ ____ ____. With these two characteristics you will have little difficulty solving problems involving friction.

oppose

impends

is taking place

Summary

Dry friction is the resistance to slip of two dry surfaces that are pressed together. Friction forces are always directed so as to oppose relative motion of the two bodies in contact and are here considered to depend only on the forces acting on the bodies and a material property, μ, the coefficient of friction. By definition, $\mu = f_{max}/N$ where f_{max} is the maximum friction force possible when the normal force, N, presses two bodies together. Also, the angle of friction, ϕ, is defined by the relation $\tan \phi = f_{max}/N = \mu$.

Equilibrium problems with friction are classified into the following three types.

1. Motion impends or is taking place at a given number of known contact surfaces.
2. Impending or actual motion exists but the exact nature of this motion and the surfaces on which the motion takes place are not known.
3. It is not known whether or not motion impends or is taking place.

Problems of type (1) are solved by writing $f = \mu N$ for all surfaces where relative motion impends or is taking place. Type (2) problems are usually solved by assuming that equilibrium exists, solving, and then checking for violations of equilibrium and friction laws. Type (3) problems require assumptions of plausible modes of motion, solution, and subsequent checking for violations of equilibrium and friction requirements.

Relations between torque and thrust on square threaded screws with friction are obtained by treating them as wedges wrapped around an axis. The relation $T_{\text{Large}} = T_{\text{Small}}\, e^{\mu\theta}$ is derived for the tension T in a flat belt wrapped around a fixed body through a contact angle θ.

Problems

(1) A vehicle, shown in Fig. P8.1, with four wheels is standing on a slope with an angle of inclination α and is to be held in position by locking all four wheels. Show that this is possible only if the coefficient of friction μ between each wheel and the ground is at least tan α. Assume that the coefficient of friction is the same for all four wheels.

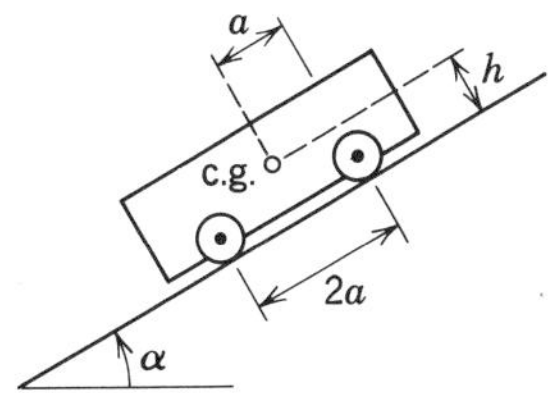

Figure P8.1

(2) What is the force F required to the wedge in Fig. P8.2 in order to prevent the body of weight Q from falling? The coefficient of friction between all surfaces is μ.

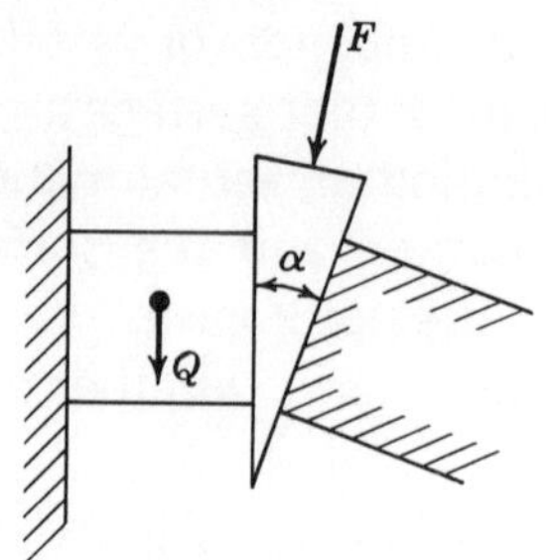

Figure P8.2

(3) What is the maximum force F a man can support when he pulls on a rope with a force P if the rope is wound 3 times around a fixed pole of diameter d. The coefficient of friction between the rope and the pole is μ.

(4) A band brake is arranged as shown in Fig. P8.3. If the cylinder rotates clockwise, what is the force on the cylinder?

(5) If the cylinder rotates counterclockwise, what is the braking force on the cylinder in Fig. 8.3?

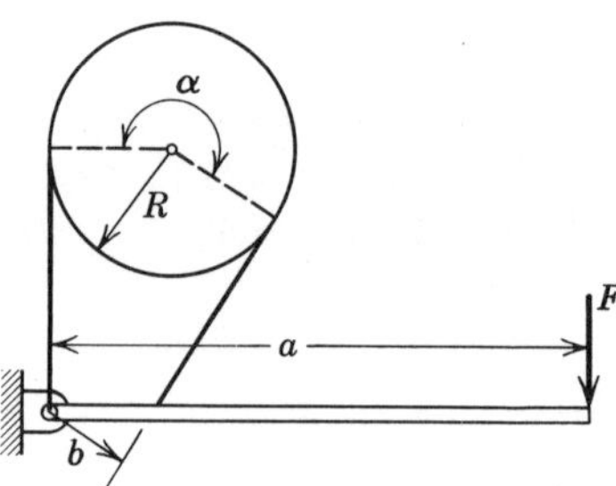

Figure P8.3

(6) A thin homogeneous rod of length ℓ and weight Q is resting in equilibrium upon a rigid cylinder of radius r as shown in Fig. P8.4. There is friction between the rod and cylinder, the coefficient of friction being μ. Suspension of a weight G at the end of the rod will cause it to roll a small distance upon the cylinder surface until a new equilibrium position is obtained. What is the largest value of G which can be applied without causing the rod to slide?

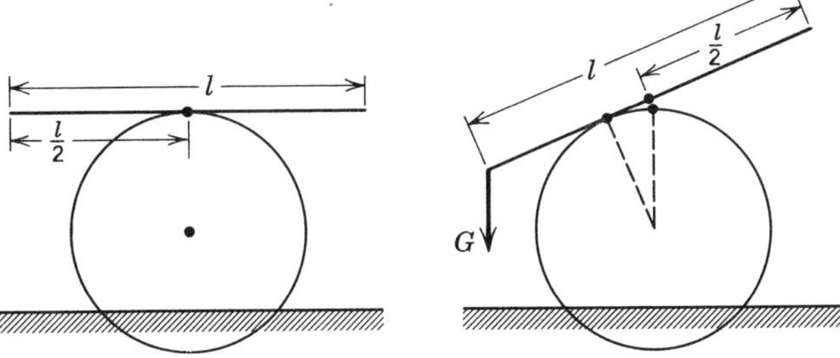

Figure P8.4

Answers to Problems

Chapter 8

(1) –

(2) $F = \frac{Q}{\mu} \sin\left(\frac{\alpha}{z}\right) - Q \cos\left(\frac{\alpha}{z}\right)$

(3) $F = P\, e^{6\pi\mu}$

(4) $f = \frac{Fa}{b}\,(e^{\mu\alpha} - 1)$

(5) $f = \frac{Fa}{b}\,(1 - e^{-\mu\alpha})$

(6) $G_{\max} = \frac{Q}{\left(\frac{1}{2r\tan^{-1}\mu} - 1\right)}$

chapter 9

Second Moments—Moments of Inertia

Objectives

In many problems of dynamics and structural analysis second moment quantities, often called moments of inertia, are present. This chapter should prepare the reader to:

1. Define regular and polar second moments of areas and second moments of masses.
2. Compute these second moments for simple geometrical shapes using direct integration.
3. Derive parallel axis theorems for area and mass second moments and apply them correctly to determine the second moments of composite shapes.
4. Show the relation between the area and mass second moments of thin plates.

9.1 SECOND MOMENTS OF AREAS

In Chapter 5, we used *first* moments of mass, volume, and area to determine centers of mass, centers of gravity, and centroids. In many problems, properties called *second moments* are important. To begin, we examine the second moment of area. Consider the area in Fig. 9.1. We define I_x, the second moment with respect to the x axis, as $\int_{\text{area}} y^2 \, dA$. Note that I_x involves an integral of y^2, where y is the distance *from the axis under consideration, (x), to the*

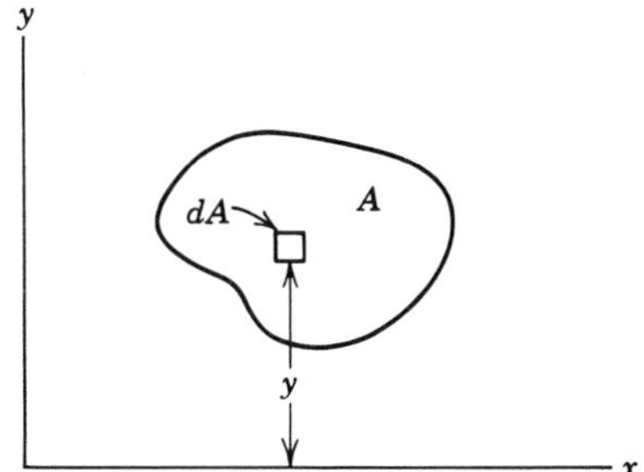

Figure 9.1

area element. Similarly, I_y is defined as $\int_{\text{area}} \_\_ dA$. Thus, *second* moments are associated with the *squares* of distances to area elements. x^2

Returning to $\int_{\text{area}} y^2\, dA = I_x$, let us calculate this quantity for the rectangular area in Fig. 9.2. To use

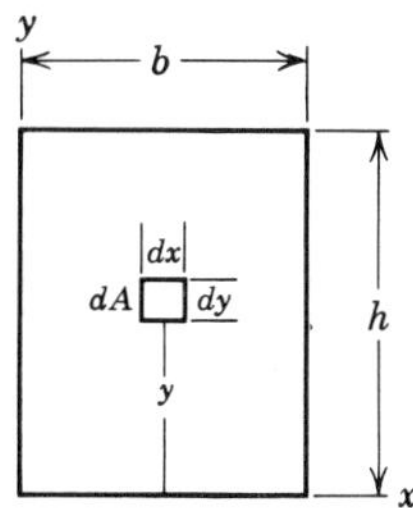

Figure 9.2

direct integration, first write $dA = dx\, dy$. The integral becomes $\int y^2\, dx\, dy$. Rewrite the integral in iterated form, putting in limits as appropriate, as $I_x =$ ________________. $\int_{y=0}^{y=h} y^2 \left[\int_{x=0}^{x=b} dx\right] dy$

Perform the integration between limits to find $I_x =$ ____. $bh^3/3$

The second moment of a rectangle with respect to an axis through its base is 1/3 the product of the base and the cube of the height. By analogy, using the symbols shown in Fig. 9.2, I_y equals (1/3) (____). hb^3

Consider the area of Fig. 9.3, bounded by the x axis, the line $x = b$ and the curve $y_c = kx^2$. The value of k can be computed by noting that $y = a$ when $x = b$.

a/b^2

Thus k equals ___.

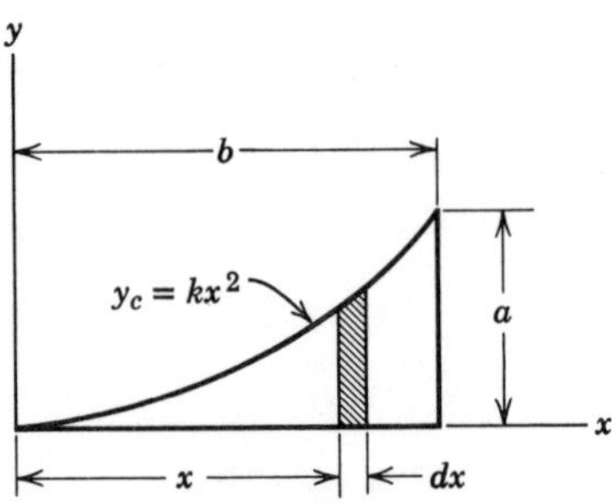

Figure 9.3

To compute I_y, we could use double integration. Instead, we write a single integral using the area element shown in Fig. 9.3. This element is $dA = y_c\, dx$. Since the entire area element has the same x coordinate, we can write $\int_{\text{area}} x^2\, dA$ as $\int_x x^2\, y_c\, dx$. Now substitute $y_c = ax^2/b^2$ into the integral and obtain an expression

ax^4/b^2

in terms of x only as $I_y = \int_x$ ___ dx.

b

The limits of integration are $x = 0$ to $x =$ ___.

$ab^3/5$

Evaluate this integral; the result is $I_y =$ ___.

Another way to determine second moments of areas is to sum the second moments of elemental areas, each of which has a *known* second moment. For example, consider again the elemental area of height y_c and width dx in Fig. 9.3. To determine I_x, *we write an expression for the second moment of the differential element and call it* dI_x. The differential element is essentially a rectangle of width dx and height y_c. The second moment of this rectangle about the x axis equals 1/3 the product of base and height

$y_c^3 dx/3$

cubed. Thus, for the differential element, $dI_x =$ ___.

To find I_x we sum by integration the contributions of all elements. Thus write $\int_{\text{area}} dI_x = \int_{\text{area}} (y_c^3/3)dx$. Insert the equation of the curve, and perform the integration

$a^3 b/21$

to find $I_x =$ ___.

We have three techniques for calculating second moments of areas. These are: double integration, single integration in which an area element with one finite dimension is used, and a process of summing the second moments of elemental areas. You should learn to choose the most efficient technique and develop facility with such calculations through practice.

Another type of second moment is the *polar second moment*. Consider Fig. 9.4. We define the

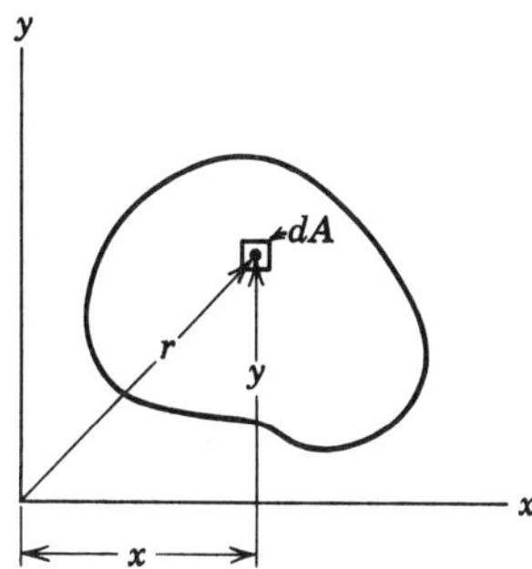

Figure 9.4

polar second moment, J_o, of this area about the point O as $\int_{\text{area}} r^2\, dA$. This integral is often calculated using polar coordinates. For example, we calculate the polar second moment of a circular area, Fig. 9.5, about its center. In polar coordinates, dA is expressed as $dA = rd\theta$ (__). *dr*

Thus for the circle, $J_o = \int_{\text{area}} r^2\, dA$ equals $\int_{\text{area}} r^3\, dr\, d\theta$.

We must integrate θ between limits O and __ and r between limits O and __. 2π R

Perform this integration to find J_o = ____ for a circle about its center. $\pi R^4/2$

We return now to Fig. 9.4 to determine a useful relationship between I_x, I_y, and J_o. Clearly, from Fig. 9.4, $x^2 + y^2 = (\_\_)^2$. r

Therefore, $J_o = \int_{\text{area}} r^2\, dA = \int_{\text{area}} (x^2 + \_\_)\, dA = I_y + \_\_$. y^2 I_x

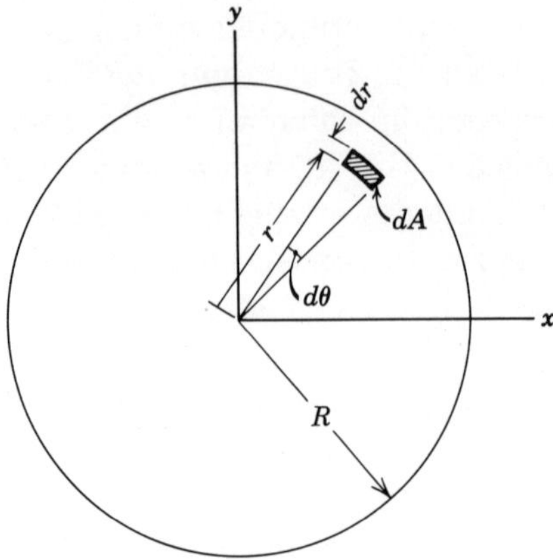

Figure 9.5

Thus we have the relationship $J_o = I_x + I_y$. We now easily find I_x or I_y for a circular area. First, due to

I_y J_o

symmetry, I_x equals __. Therefore, $2I_x =$ __.

Since we know J_o from previous calculation, I_x

$J_o/2 = \pi R^4/4$

$= I_y =$ __________.

Another useful property of areas is the *radius of gyration*. Consider an arbitrary area, as in Fig. 9.4. We define the distance k (called the radius of gyration) associated with the second moment of this area about the x axis as follows: $I_x = k_x^2 A$ where A is the total area. Similarly, we define the polar radius of gyration, k_o, by the equation $J_o = k_o^2 A$. From the relation

k_o^2

between I_x, I_y, and J_o, we note that $k_x^2 + k_y^2 =$ __.

Now let's derive a relation between the second moments of an area with respect to two parallel axes, Fig. 9.6. Remember, $I_x = \int_{\text{area}} y^2 \, dA$. Express y as the sum of the fixed distance d between the x and x' axes, and the variable distance y' from the x' axis to dA. Thus

y'

$y = d +$ __.

Next rewrite $\int_{\text{area}} y^2 \, dA$ as $\int_{\text{area}} (y' + d)^2 \, dA$. The squared term when expanded yields three integrals and the

y'^2 $2y'd$

equation becomes $I_x = \int_{\text{area}} (\_\_) \, dA + \int_{\text{area}} (\_\_\_) \, dA +$

d^2

$\int_{\text{area}} (\_\_) \, dA$

Now, examine the integrals individually. The first

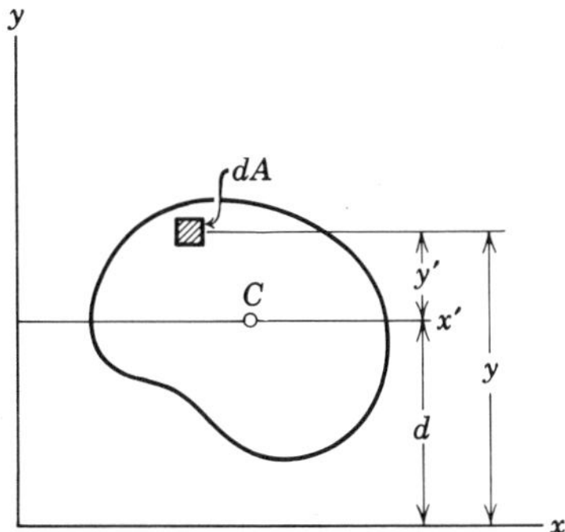

Figure 9.6

is clearly the second moment of the area with respect to the ___ axis. x'

It is called $I_{x'}$. The third integral is a constant, d^2, multiplied by the integral of dA, and equals d^2A. The second integral $2d \int_{\text{area}} y'\, dA$, is a constant multiplied by the *first* moment with respect to the x' axis. We now impose an important restriction by *limiting use of this formula to an axis x′ that passes through the centroid, C, of the area.* With this restriction, the first moment with respect to the x' axis must equal ___. Therefore, the second integral equals ___. zero zero

Since x' passes through C, $I_{x'}$ becomes $I_{x'_c}$, the second integral is zero, and we obtain finally $I_x = I_{x'_c} + d^2A$. This relation, called the *parallel axis theorem*, states that the second moment with respect to an axis, x, equals the second moment with respect to a parallel centroidal axis, x_c', plus a transfer term equal to the product of the area and the square of the distance between the x and x' axes.

Since areas of interest often consist of a collection of common geometric shapes, we next describe a method, called the *composite area technique*, that simplifies the determination of their second moments. We shall calculate I_x for the *T*-shaped area of Fig. 9.7. First, picture this *T*-section as divided into rectangles, 2″ high by 12″ wide (D) and 6″ high by 2″ wide (B). We now use the relation for the second moment of a rectangle about an axis through its base ($I = bh^3/3$) and the parallel axis theorem to deter-

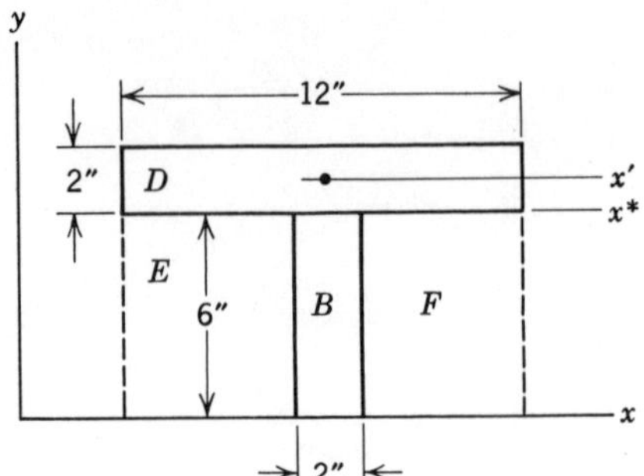

Figure 9.7

mine the second moment of the T-section with respect to the x axis. For rectangle B, $I_{x-B} = (1/3)\ (2'')$
6" 144 $(\_\_)^3$ which equals ___ in.^{4}

We now seek the second moment of D with respect to the x axis. To find I_{x-D} using the parallel axis theorem, we need $I_{x'_c-D}$. The formula $I = (1/3)\,bh^3$ gives I_{x^*-D}. The parallel axis theorem, $I_{x^*-D} = I_{x'_c-D} + d^2A$, is used to find $I_{x'_c-D}$. Thus $I_{x'_c-D} = I_{x^*-D} - d^2A$
1" 24 in.^{2} where d equals ___ and A equals _____

8 in.^{4} Evaluate the expression to find $I_{x'_c-D}$ equals ____

With $I_{x'_c-D}$ known, we use the parallel axis formula again to determine I_{x-D}. Thus $I_{x-D} = I_{x'_c-D} + d^2A = 8$
7" $+ (\_\_)^2\ (24)$.

1184 in.^{4} Now, find the numerical value I_{x-D} equals _____

The total second moment of the T-section would
1328 in.^{4} thus be $I_{x-D} + I_{x-B} =$ ______

It is important to note that we *could not* use the parallel axis theorem to determine I_x directly from I_{x^*} but instead *had to determine* $I_{x'_c}$ *first. Remember,* when using this theorem in the final form given, you
centroid *must* have one of the axes through the _____ of the area.

Another way to solve this problem is to determine the *difference* between the moment of inertia of the rectangle of width 12" and height 8" (Fig. 9.7) and the combined moments of the two areas, E and F. As an exercise, solve the problem this way.

9.2 SECOND MOMENTS OF MASSES —MOMENTS OF INERTIA

In the study of the dynamics of rigid bodies, the second moment of mass or the *mass moment of inertia* is important. We define the mass moment of inertia of a body with respect to an axis as $I = \int_{\text{mass}} r^2\,dm$, where r is the distance from the mass element, dm, to the axis. We shall first develop a parallel axis theorem for the mass moment of inertia. Consider the arbitrary body of Fig. 9.8. We examine the mo-

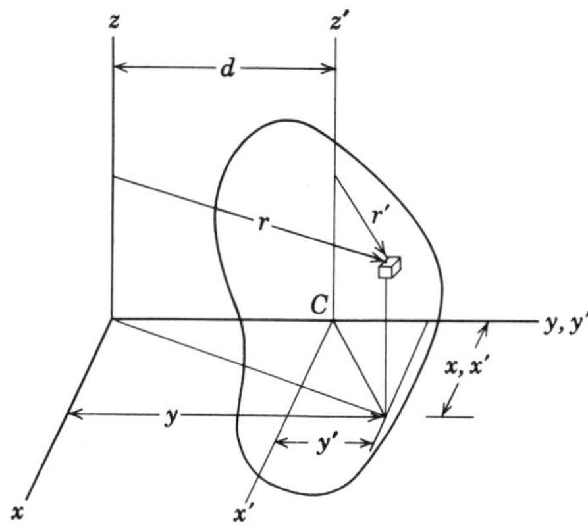

Figure 9.8

ment of inertia of the body about the z axis and seek a relationship between I_z and $I_{z'}$, where the z' axis is parallel to the z axis and intersects C (the center of mass of the body). The coordinate system is oriented so that *the y axis passes through point C*. In this case, using the notation of Fig. 9.8, we write $I_z = \int_{\text{mass}} r^2\,dm$ and $I_{z'} =$ ______. $\int_{\text{mass}} r'^2\,dm$

Using the relationships $x^2 + y^2 = r^2$ and $x'^2 + y'^2 = r'^2$, rewrite the two integrals as $I_z = \int_{\text{mass}} (x^2 + y^2)\,dm$ and $I_{z'} = \int_{\text{mass}} (\_\_\_\_\_\_)\,dm$. $x'^2 + y'^2$

d Recognize also that $x = x'$ and $y = \_\_ + y'$.

Thus we express I_z and $I_{z'}$ as $I_z = \int_{\text{mass}} x^2 dm + \int_{\text{mass}} (d + y')^2 dm$ and $I_{z'} = \int_{\text{mass}} x^2 dm + \int_{\text{mass}} y'^2 dm$. Comparing these expressions, we see that we can write $I_z = I_{z'}$

y' $+ \int_{\text{mass}} d^2 dm + \int_{\text{mass}} 2d\,(\_\_)dm$.

Since d is a fixed distance, this expression becomes $I_z = I_{z'} + d^2 \int_{\text{mass}} dm + 2d \int_{\text{mass}} y'dm$. Clearly $\int_{\text{mass}} dm$ is

Mass the total ____ of the system (M).

The origin of the y' axis is point C, the center of

zero mass of the body. By definition, $\int_{\text{mass}} y'dm = \_\_\_\_$.

$I_{z'}$ d^2M Thus the parallel axis theorem is $I_z = \_\_ + \_\_\_\_$.

Note that the theorem has been derived *specifically for the z′ axis passing through the center of mass.*

Let's compute some mass moments of inertia, using the parallel axis theorem as needed. We seek the mass moment of inertia of the slender rod of Fig. 9.9 with respect to the x axis. The rod has cross-

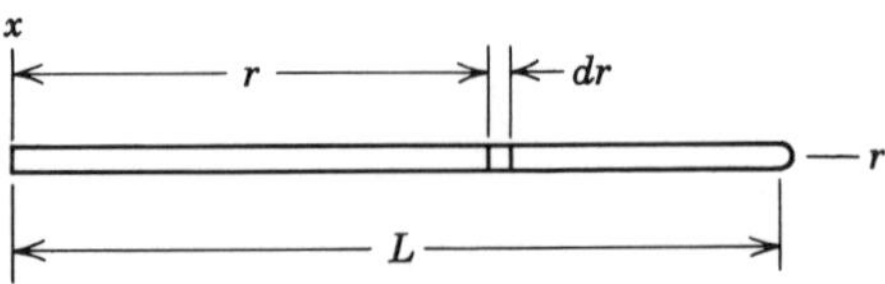

Figure 9.9

sectional area A and mass density ρ. We express the distance from the x axis to dm as r and write dm

$r^2\rho A\,dr$ equals $\rho A\,dr$. Thus I_x equals $\int_0^L \_\_\_\_$.

With A and ρ constant, perform the integration

$A\rho L^3/3$ to obtain $I_x = A\rho r^3/3\Big|_0^L$ which equals ____.

Since the total mass of the rod is ρAL, we can rewrite the expression for I_x in terms of M, the total

$ML^2/3$ mass of the rod, as $I_x = \_\_\_\_$.

$I_{x'_c}$, the mass moment of inertia of this rod with respect to an axis through its center of mass ($r = L/2$), is easily determined using the parallel axis theorem.

$ML^2/4$ To do so, we write $I_x = I_{x'_c} + d^2M = I_{x'_c} + \_\_\_\_$.

$ML^2/12$ The result is $I_{x'_c} = \_\_\_\_$.

Next, consider the mass moment of inertia of

thin plates (Fig. 9.10). The plate *is considered sufficiently thin* that it may be treated as lying in the y-z

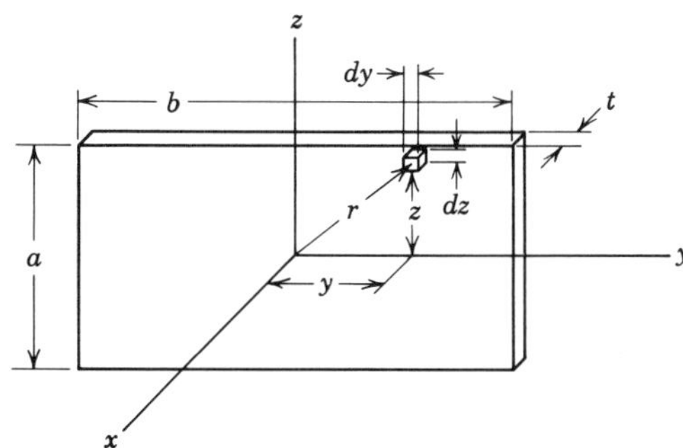

Figure 9.10

plane. The material of the plate has mass density ρ, and dm equals $\rho t dA$. Write $dA = dydz$ and use the notation of Fig. 9.10, to obtain $I_x = \int_{\text{area}} (\_\_)^2 (\_\_)(\_\_) dA$. $r \quad \rho \quad t$

With ρ and t constant, this becomes $\rho t \int_{\text{area}} r^2 dA$. Note that $\int_{\text{area}} r^2 dA$ is the polar second moment of the surface area of the plate. This integral is evaluated using the relation between ordinary second moments and polar moments of areas, $\int r^2 dA = \int y^2 dA + \int \_\_ dA$. z^2

Thus the mass moment of the plate, I_x, equals $\rho t \int_{\text{area}} y^2 dA + \rho t \int_{\text{area}} z^2 dA$, that is, $I_x = I_z + \_\_$. I_y

This relation is true for *thin plates* but not for three-dimensional solid bodies in general. The fact that $I_x = I_y + I_z$ simplifies many computations. For example, we remember (or can easily calculate) that for a rectangular area with dimensions a and b as shown, $\int_{\text{area}} z^2 dA = (1/12)ba^3$. Thus, for the plate, the mass moment of inertia I_y equals $(\_\_)(\_\_)(1/12)ba^3$. $\rho \quad t$

Similarly, I_z equals $(1/12)\rho t (\_\_)(\_\_)^3$. $a \quad b$

Thus $I_x = \rho \frac{t}{12}(ab^3 + ba^3)$. We can simplify further since the total mass of the plate equals the product of its area, thickness and density, $M = abt\rho$. With this

b^2 a^2 relation we obtain $I_x = \frac{M}{12}(\_\_ + \_\_)$.

Another example, is the pyramid in Fig. 9.11.

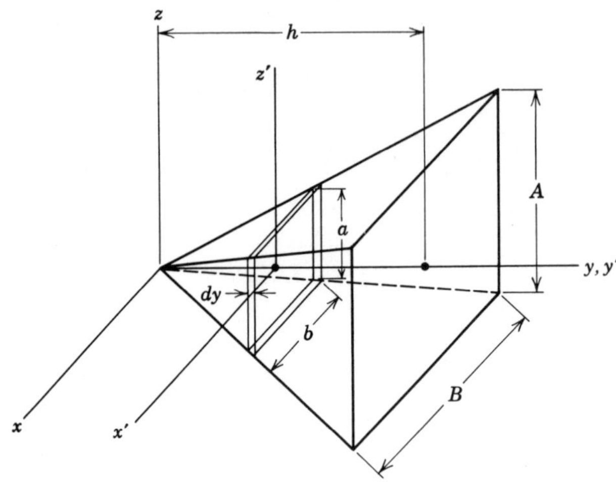

Figure 9.11

Rather than use multiple integration, let us imagine that the pyramid is divided into thin rectangular plates and add the mass moments of these plates to find the mass moment of inertia of the entire pyramid.

We seek I_x. To find the mass moment of inertia I_x of a thin plate element of width b, height a, and thickness dy, we express $I_{x'}$ for the plate element and use the parallel axis theorem. As determined in a previous example, $I_{x'}$ of a thin plate element equals the product of $(1/12)\rho$ · (thickness) · (base) · (height

dy cubed). Thus $I_{x'}$ for our element is $I_{x'} = \frac{1}{12}\rho \cdot (\_\_)$

b $(\_\_)a^3$.

The parallel axis theorem applied to the plate, becoming $I_x = \frac{1}{12}\rho b a^3 \, dy + y^2 M$ when M is the mass

$\rho ab\,dy$ of the plate and equals ____.

This thin plate is a differential element of the entire pyramid. We have an expression for the moment of inertia of the thin plate with respect to the x axis. This represents, however, only a differential part, dI_x, of the mass moment of inertia of the entire pyramid. Our remaining task is to sum the contributions of all plate elements by evaluating the integral $I_x = \int_{\text{mass}} dI_x = \int_0^h \left(\frac{1}{12}\rho b a^3 + y^2 \rho ab\right) dy$. The dependence of a and b on y is determined from similar triangles. For example, $a/A = y/h$. Similarly, b equals ___. By/h

The expressions for a and b just obtained are substituted into the $\int dI_x$. Our final expression is

$$\int dI_x = (1/12)\int_0^h [\_\_\_\_\_\_\_\_]\, dy + \int_0^h y^2\, [\_\_\_\_\_\_\_\_]\, dy.$$

$\rho(by/h)(Ay)^3/h$
$\rho(Ay/h)(By/h)$

Integrate this expression between the limits $y = 0$ and $y = h$ (on scratch paper). The final result for the pyramid is I_x equals ________. $\rho ABh\left(\frac{A^2}{60} + \frac{h^2}{5}\right)$

For practice, repeat this process to find I_z. Done properly you will find that I_z equals ________. $\rho ABh\left(\frac{B^2}{60} + \frac{h^2}{5}\right)$

Our final topic is the calculation of mass moments of inertia for solid bodies composed of standard geometrical shapes by use of a composite body approach. Consider a thin rod attached to the center of a disk, Fig. 9.12. The moment of inertia of the system about the x axis is readily obtained by the composite body approach. First the contribution of the rod is $I_{x-\text{rod}} = M_{\text{rod}}L^2/3$ (determined previously). The contribution of the disk is determined using the parallel axis theorem. Recall (or calculate, if necessary) that $I_{x'}$ for a disk is $I_{x'} = (1/4)M_{\text{disk}}R^2$. The parallel axis theorem indicates that the moment of inertia

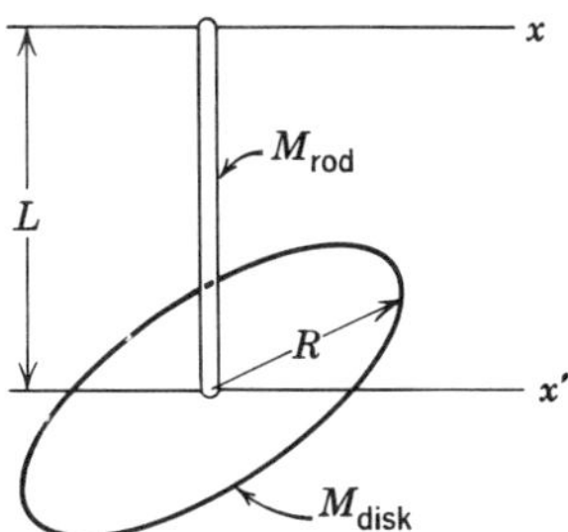

Figure 9.12

of the disk about the x axis is $I_{x-\text{disk}} = (1/4)M_{\text{disk}}R^2 + \_\_^2 M_{\text{disk}}$.

L

Combining the expressions for the rod and the disk, we obtain $I_{x\,(\text{rod}+\text{disk})} =$ ________________.

$M_{\text{rod}}\dfrac{L^2}{3} + M_{\text{disk}}\left(\dfrac{R^2}{4} + L^2\right)$

Summary

The second moment I_x of an area in the x-y plane about the x axis is defined as

$$I_x = \int_A y^2\,dA$$

Similarly I_y, about the y axis, equals $\int_A x^2 dA$. The polar second moment J_o about some point O in the plane is defined as

$$J_o = \int_A r^2\,dA.$$

Evaluation of these quantities involves a straightforward multiple integration.

The parallel axis theorem $I_x = I_{x'_c} + d^2 A$, where x'_c is an axis through the centroid parallel to x, d is the distance between x and x'_c, and A is the total area, is used to determine the second moment of composite areas. The technique involves decomposition of areas into simple geometrical elements and the summation of the second moments of these elements.

The second moment of mass or moment of inertia, I, with respect to an axis, is defined as

$$I = \int_{\text{mass}} r^2\,dm.$$

Such integrals are evaluated by expressing dm as ρdV and integrating over the volume of the body. As with areas, we have a mass parallel axis theorem $I_z = I_{z'} + d^2 M$ where the z' axis is parallel to z and passes through the center of mass of the body, d is the distance between z and z', and where M is the total mass of the body. This theorem is used to determine the moment of inertia of bodies that can be subdivided into simple geometrical shapes using the composite body approach.

Problems

(1) By integration find the moment of inertia I_y of the rectangle in Fig. P9.1. Check the result by using the parallel axis theorem.

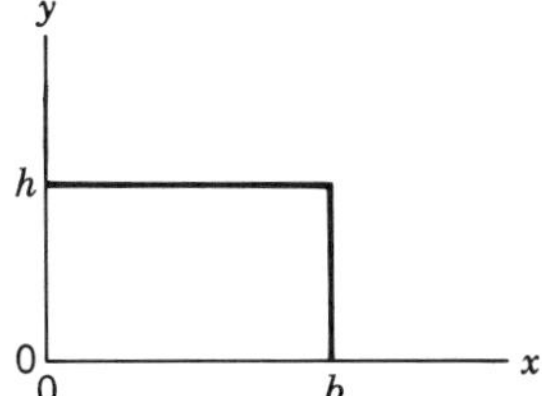

Figure P9.1

(2) Determine the moment of inertia about the $x'x'$ axis of the cross-sectional area shown in Fig. P9.2.

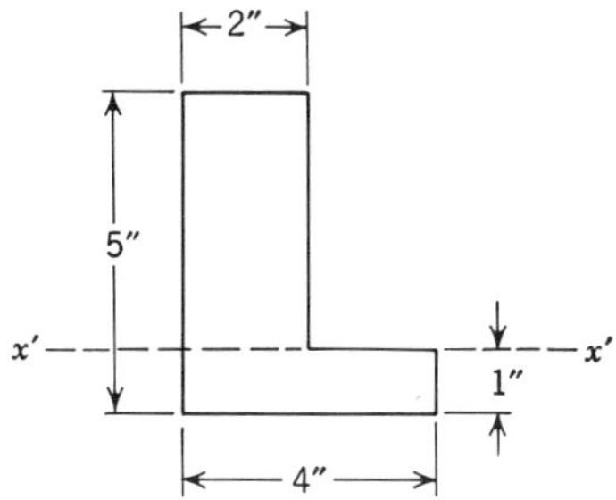

Figure P9.2

(3) Calculate the magnitude of $\bar{\mathbf{F}}$ per unit width necessary for equilibrium of the gate in Fig P9.3.

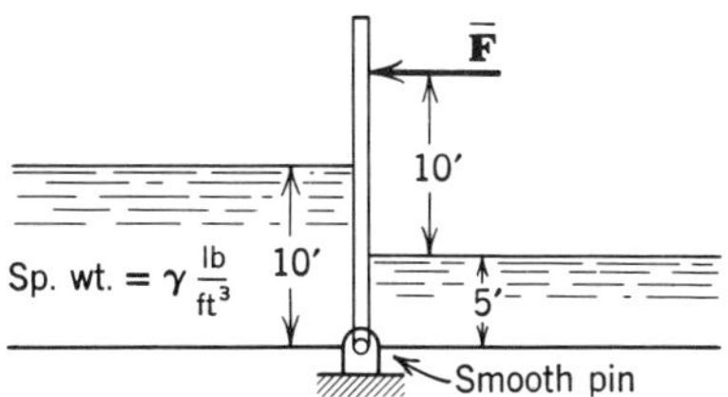

Figure P9.3

(4) Find the area moment of inertia of the channel cross section shown in Fig. P9.4 (with respect to the $x'\ x'$ axis).

(5) If the base of the channel in Fig. P9.4 is made of steel, $\rho_s = 491$ lb/ft³, and the legs are aluminum, $\rho_A = 175$ lb/ft³, find the mass moment of inertia about the base $x'x'$ (channel length

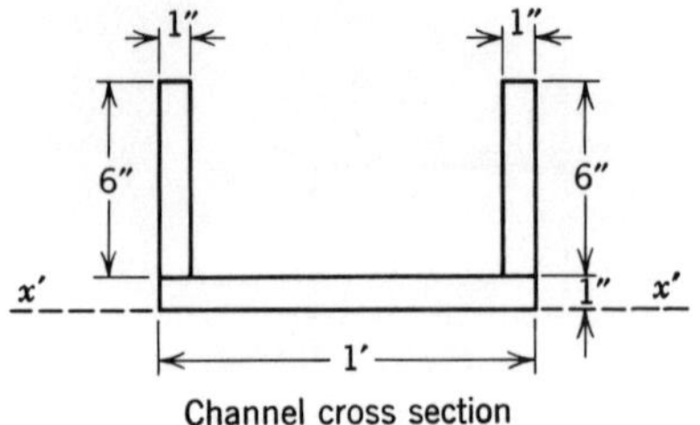

Figure P9.4

= 10′). Is the mass moment of inertia directly proportional to the area moment of inertia found in (4)?

(6) By integration, calculate the polar moment of inertia about point O for the segment of a circle shown in Fig. P9.5.

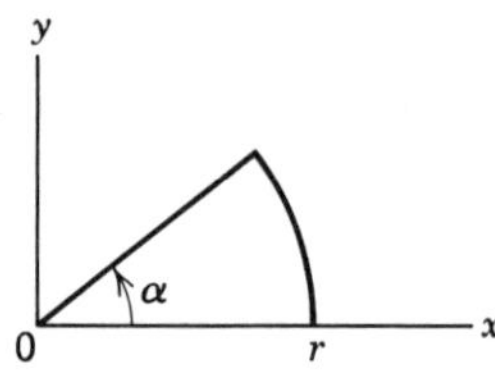

Figure P9.5

Answers to Problems

Chapter 9

(1) $I_y = \dfrac{b^3 h}{3}$

(2) $I_{r'} = 44$ in.^{4}

(3) $F = 9.7\,\gamma$ lb/ft width

(4) $I_{x'} = 154$ in.^{4}

(5) $I_{x'x'} = 19.4$ lb-ft^2
No

(6) $I_o = \dfrac{\alpha r^4}{4}$

chapter 10

Virtual Work

Objectives

The principle of virtual work is a powerful alternative tool for solving problems of equilibrium. This introduction to the principle should make it possible for the student to:

1. Define a virtual displacement and virtual work.
2. State the principle of virtual work as it applies to rigid bodies and connected systems of rigid bodies without friction.
3. Apply the principle to solve simple equilibrium problems involving forces and reactions on rigid bodies and connected systems of rigid bodies without friction.

Our final topic is the principle of virtual work, which is a powerful tool of mechanics with application far beyond statics. We limit ourselves to the principle as applied to rigid bodies in equilibrium.

Before stating the principle, we need several definitions. We first define the work of a force, $\bar{\mathbf{F}}$, acting through an infinitesimal displacement, $\overline{\mathbf{dr}}$, where $\overline{\mathbf{dr}}$ is a vector from the initial to the final position of the point of application of the force. By definition, the work dU done by the force $\bar{\mathbf{F}}$ acting through $\overline{\mathbf{dr}}$ equals $\bar{\mathbf{F}} \cdot \overline{\mathbf{dr}}$, that is, $dU = \bar{\mathbf{F}} \cdot \overline{\mathbf{dr}}$. Thus, work is a ____________ quantity and may be positive or negative. vector/scalar

scalar

From the definition, note that if $\bar{\mathbf{F}}$ and $\overline{\mathbf{dr}}$ are parallel, $dU = F\,dr$, and if $\bar{\mathbf{F}}$ and $\overline{\mathbf{dr}}$ are perpendicular, $dU =$ ____.

zero

Also, if the point of application of the force does not move, that is, if $\overline{\mathbf{dr}} = 0$, then $dU =$ ___.

zero

We also need to define the work done by a couple acting through an infinitesimal angular displacement $\overline{\mathbf{d\theta}}$, where $\overline{\mathbf{d\theta}}$ is the vector representing the difference between the final and initial angular positions of the couple. By definition, $dU = \overline{\mathbf{M}} \cdot \overline{\mathbf{d\theta}}$. This can be derived from the relation $dU = \overline{\mathbf{F}} \cdot \overline{\mathbf{dr}}$ by considering the forces of the couple $\overline{\mathbf{M}}$. Thus the work of a couple is zero if the angular displacement $\overline{\mathbf{d\theta}}$ is zero or if it is ________________ to $\overline{\mathbf{M}}$.
parallel/perpendicular

perpendicular

A *virtual displacement* is defined as an imaginary infinitesimal displacement (linear or angular) consistent with the constraints of a system. *Virtual work* is defined as the work done by the external forces acting on a system that undergoes a virtual displacement.

We now state the *principle of virtual work. For a rigid body in equilibrium, the total virtual work of the external forces acting is zero for any virtual displacement.*

We demonstrate the use of this principle through an example. Figure 10.1 shows a ladder on rollers

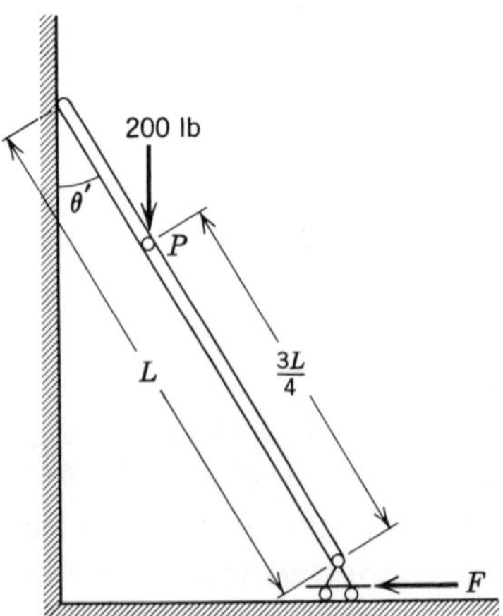

Figure 10.1

leaning against a smooth wall. We seek the force F that must be applied at the foot to maintain equilibrium if a 200-lb man stands at *P*. We first draw a free body diagram of the ladder, Fig. 10.2. Since the

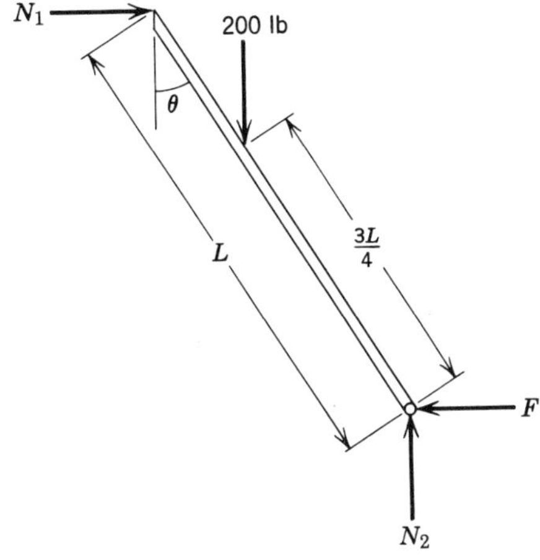

Figure 10.2

wall is smooth, the reaction force at the wall is ________________ to the wall. perpendicular

parallel/perpendicular

We now choose our virtual displacement to be a displacement δx of the foot of the ladder to the right and an associated rotation of the ladder, $\delta\theta$. Of course, during this virtual displacement, the top of the ladder moves ________. down

up/down

Figure 10.3 shows the ladder in its displaced position. During the virtual displacement, the forces N_1 and N_2 do no work, since they move ________ to their lines of action. perpendicular

The virtual work of F is $dU_F = -F$ (__). δx

The virtual work of the 200-lb force is the product of the force magnitude and the displacement of the force parallel to its line of action. Thus the virtual work of the 200-lb force is $dU_{200} = 200\ (\delta y) = 200 \left(\frac{3L}{4}\,\delta\theta\right)$ (___). $\sin\theta$

However, δx and $\delta\theta$ are related, $\delta x = L\delta\theta\cos\theta$ (Fig. 10.3). We thus write $dU_{200} = 200\left(\frac{3L}{4}\right)\left(\frac{\delta x}{L\cos\theta}\right)$.

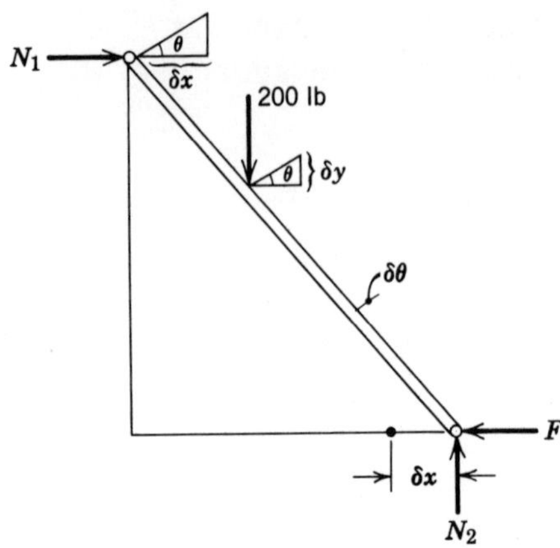

Figure 10.3

$\tan\theta$

$(\sin\theta)$ which equals $200\left(\frac{3\delta x}{4}\right)\cdot(\_\_)$.

We now write the principle of virtual work: $dU_F + dU_{200} = 0$ or $-F\,\delta x + 150\delta x \tan\theta = 0$.

$150\tan\theta$ lb

Divide by the factor δx and solve to find $F = \_\_\_\_\_$.

You can check this result by using equilibrium equations for the free body shown in Fig. 10.2.

The principle of virtual work is particularly useful in analyzing connected systems of rigid bodies. We limit consideration to systems without friction. *For a frictionless, connected system of rigid bodies in equilibrium, the total virtual work of the external forces acting is zero for any virtual displacement.*

displacement

To demonstrate, we shall determine the magnitude of the force P necessary for equilibrium of the pin-connected linkage of Fig. 10.4. We choose as our virtual displacement an angular displacement $\delta\theta$ of all members as shown in Fig. 10.5. We calculate the virtual work. The reaction force at A does no work, since there is no _____ of point A.

$2\delta\theta\sin 45°$

The 100-lb forces at B and G have equal virtual work. The virtual work of each is $-100\ \delta x$. Since $\delta x = 2\delta\theta \sin 45°$, the total virtual work of the two 100-lb forces is $dU_{100} = -2(100)\ (\_\_\_\_\_)$.

$\delta\theta$

Both of the 300 ft-lb couples are rotated through an angle $\delta\theta$. Their combined virtual work is $dU_{300} = -2(300)\ (\_\_)$.

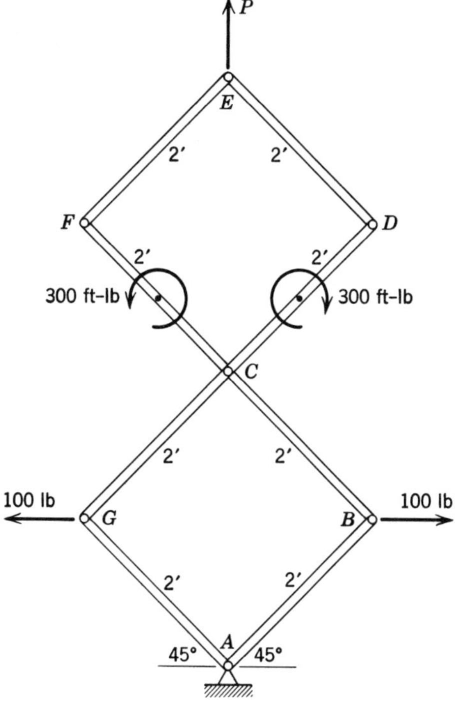

Figure 10.4

From Fig. 10.5 we see that the vertical component of the virtual displacement of B (δy) equals 2(__) sin 45°. $\delta\theta$

The vertical components of the virtual displacements of points C, D, and E equal $2\delta y$, $3\delta y$ and $4\delta y$, respectively. Therefore, the virtual work of P is $dU_p = 4P\,\delta y = 4P$ (________). $2\delta\theta \sin 45°$

The principle of virtual work states that $dU_{100} + dU_{300} + dU_p = 0$. Substitute in the expressions for these quantities and obtain the equation $(-400)\,\delta\theta \sin 45° -$ (_____) + (__________) = 0. $600\delta\theta$ $8P\,\delta\theta \sin 45°$

Divide by $\delta\theta$ and solve to determine that P = __________________ lb. $50 + 150/\sqrt{2} = 156$

Thus we have obtained P without examining the members of the system individually.

The principle of virtual work can also be used to determine reaction forces at fixed supports. This is done by imagining that the support undergoes a virtual displacement in the direction of the force component sought. For example, we seek the hori-

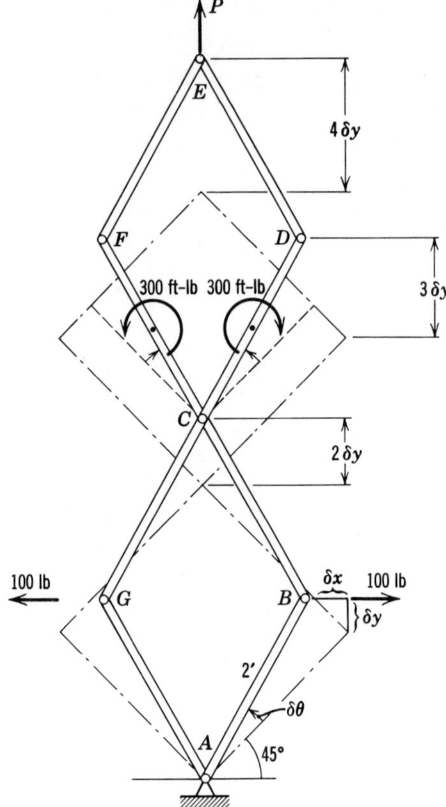

Figure 10.5

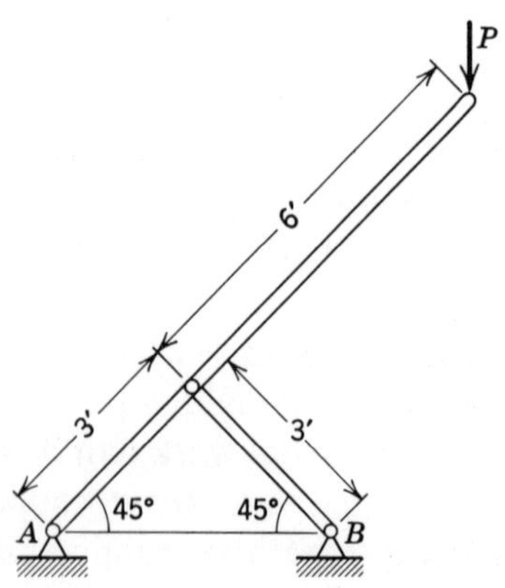

Figure 10.6

zontal reaction component B_x of the frame of Fig. 10.6. It can be determined using virtual work by imagining the system to be as shown in Fig. 10.7. Allow a virtual displacement of B and use virtual work to determine B_x. You should obtain B_x = ___ lb. 3P/2

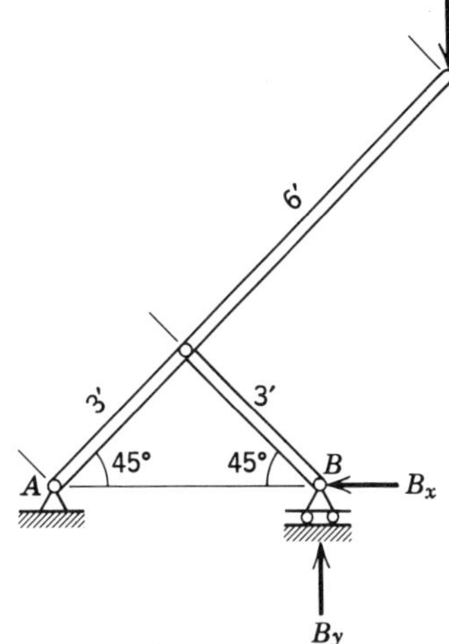

Figure 10.7

Summary

A virtual displacement is defined as an imaginary infinitesimal displacement (linear or angular) consistent with the constraints of a system. Virtual work is defined as the work done by the external forces on a system during a virtual displacement. The principle of virtual work states that, for a rigid body or frictionless collection of rigid bodies, the virtual work is zero for any virtual displacement. The principle is particularly useful here for complex connected systems of rigid bodies. A similar principle provides a powerful tool applicable to a wide variety of more advanced problems involving deformable bodies.

Problems

(1) Using the method of virtual work, write the equations of equilibrium for the statically indeterminate, weightless beam of Fig. P10.1.

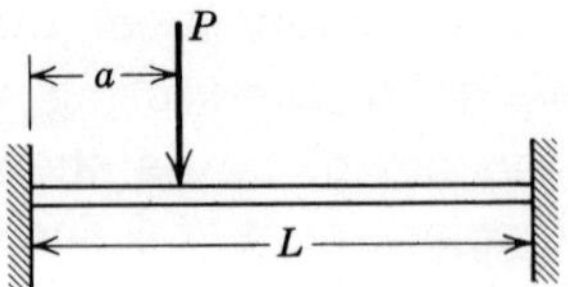

Figure P10.1

(2) Find the angle α for equilibrium of the system shown in Fig. P10.2.

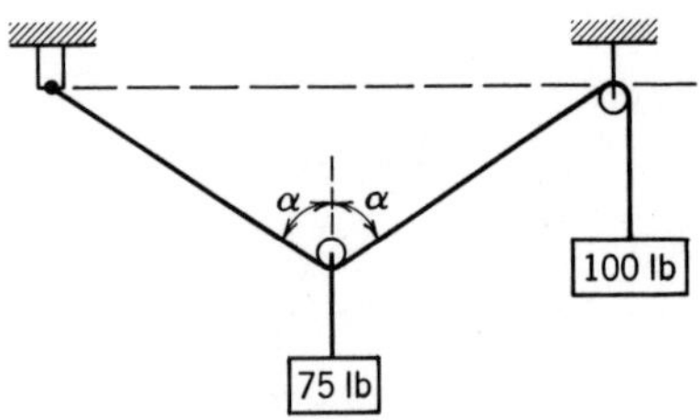

Figure P10.2

(3) When the rod of Fig. P10.3 is vertical, the spring is elongated by an amount a. Determine the angle θ for equilibrium.

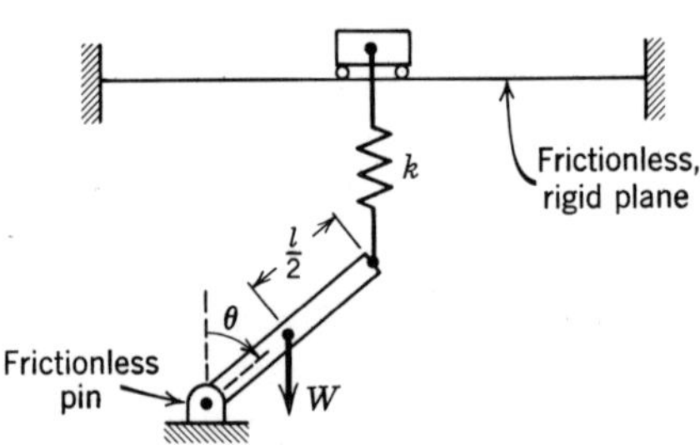

Figure P10.3

Answers to Problems

Chapter 10

(1) $-R_{Rx} + R_{Lx} = 0$
$-R_{RY} + P - R_{Ly} = 0$
$-M_R + P \cdot Q - R_{Ly}L + M_L = 0$

(2) $\alpha = 68°$

(3) $\theta = \cos^{-1}\left(1 + \frac{a}{\ell} - \frac{W}{2K\ell}\right)$

Index

Angles between lines, 17
Area, by multiple integration, 62

Beams, 107
  load-shear relation, 110
  shear-moment relation, 110
Bending Moment, 108

Cables, flexible, 116
  loaded at discrete points, 116
  parabolic, 122
  plane, 116
  sag, 118
  tension, 119
  vertical distributed loads, 120
Center of gravity of solid bodies, 59
Centroid, 57
  of an area, 61
  of a volume, 61
Chains, 116
Composite area technique, 163
Composite systems, approach to, 66
  division of, 86
  equilibrium, 84
Concentrated force, 112
Continuous beams, 87
Coordinates, orthogonal, 11
Couple, 34

Differential elements, 51
Distributed force system, resultant of, 53

Equilibrium, 2, 72
  body acted on by three forces, 83
  equations of, 78
  independent equations available, 81
  of individual joints, 100
  of a particle, 73
  requirements for a rigid body, 75
  two-dimensional, 75

Forces, concurrent, 42
  reaction, 76
  resultant, 42
  within continuous members, 87
Force system, concurrent, 43
  coplanar, 43
  distributed, 57
  equivalent, 40
  external effects, 39
  general, 39
  parallel, 46
  special, 42
Frames, 93, 104
  just rigid, 107
  plane, 104
  statically indeterminate, 107
Free body diagrams, 76
Friction, 129
  basic nature of, 130
  belt or cable, 150
  coefficient of kinetic, 132
  coefficient of static, 131
  contact area, 133
  dry sliding, 129
  fluid, 129
  general approach to problems of, 134
  internal, 129
  limiting value, 131
  in machine elements, 146
  rolling, 129

Gravity, center of, 57
  forces on a collection of particles, 57
Gyration, radius of, 162

Idealizations, 1

Joints, equilibrium, 98
  special loading conditions for, 100

Laws, basic, 1
Length, 2

Mass, definition, 2
  center of, 57, 60
Mass moment of inertia, 165
Maxwell Diagram, 104
Method of joints, 100
Method of sections, 104
Moment, 24
  curve, 114
  diagram, 115
  of a force about a line, 28
  of a force about a point, 24
  of inertia, 158, 166
  of inertia, thin plates, 166–167
  jump in the, 111
  sign convention, 109

Newton, law of gravitation, 3
  laws of motion, 3

Orthogonal projections, 17

Parallel axis theorem, 163
Particle, 2, 74
Point force, 2
Polar second moment, 161
Pressure, center of, 51
Procedure for the solution of equilibrium problems, 81

Radius of gyration, 162
Resolution of vectors, 10
Resultant, of a force system, 41
  general method for finding, 46
Rigid body, 2

Sag, 118
Scalar product of two vectors, 15
Scalar quantities, 7
Scalar triple product, 30
Second moments, 158
  of areas, 158
  of areas, techniques for calculating, 160
  of masses, 165
  polar, 161
Shear, 108
  curve, 112
  diagram, 113
  jump in the, 110
  sign convention, 109
Sign convention for internal shear and moment, 109
Square threaded screw, 148
Statically indeterminate, 81
Structural analysis, 93
Structural members, internal forces in, 107
Symmetry properties of areas and volumes, 64

Time, 2
Transforming vectors, 13
Truss, 93
  plane simple, 94
  just rigid, 95
  section of, 102
  simple, 94
  space, 101, 102
Two force member, 82

Units, changing, 4

Vector, addition, associative property, 10
  equality, 10
  magnitude of, 8
  multiplication by a scalar, 9
  orientation of, 8
  product of two vectors, 22
  product operation in component form, 23
  quantities, 7
  representation, 14
  unit, 12
Virtual displacement, 174
Virtual work, 173, 174
  connected system of rigid bodies, 176
  principle of, 174
Volume by integration, 64

Wedge, 146
Weight, 2